88 Advances in Polymer Science

Speciality Polymers/ Polymer Physics

With Contributions by
Yu. K. Godovsky, K. Horie, A. Kaneda,
N. Kinjo, L. F. Kosyanchuk, Yu. S. Lipatov,
T. E. Lipatova, I. Mita, K. Nishi,
M. Ogata, V. S. Papkov

With 102 Figures and 21 Tables

Springer-Verlag Berlin Heidelberg NewYork
London Paris Tokyo

ISBN-3-540-50472-9 Springer-Verlag Berlin Heidelberg New York
ISBN-0-387-50472-9 Springer-Verlag New York Berlin Heidelberg

Library of Congress Catalog Card Number 61-642

This work is subject to copyright. All rights are reserved, whether the whole or part of the material is concerned, specifically the rights of translation, reprinting, reuse of illustrations, recitation, broadcasting, reproduction on microfilms or in other ways, and storage in data banks. Duplication of this publication or parts thereof is only permitted under the provisions of the German Copyright Law of September 9, 1965, in its version of June 24, 1985, and a copyright free must always be paid. Violations fall under the prosecution act of the German Copyright Law.

© Springer-Verlag Berlin Heidelberg 1989
Printed in Germany

The use of registered names, trademarks, etc. in this publication does not imply, even in the absence of a specific statement, that such names are exempt from the relevant protective laws and regulations and therefore free for general use.

Bookbinding: Lüderitz & Bauer, Berlin
2152/3020-543210 — Printed on acid-free paper

Editors

Prof. Henri Benoit, CNRS, Centre de Recherches sur les Macromolecules, 6, rue Boussingault, 67083 Strasbourg Cedex, France

Prof. Hans-Joachim Cantow, Institut für Makromolekulare Chemie der Universität, Stefan-Meier-Str. 31, 7800 Freiburg i. Br., FRG

Prof. Karel Dušek, Institute of Macromolecular Chemistry, Czechoslovak Academy of Sciences, 16206 Prague 616, ČSSR

Prof. Hiroshi Fujita, 35 Chimotakedono-cho, Shichiku, Kita-ku, Kyoto 603, Japan

Prof. Gisela Henrici-Olivé, Chemical Department, University of California, San Diego, La Jolla, CA 92037, U.S.A.

Prof. Dr. habil Günter Heublein, Sektion Chemie, Friedrich-Schiller-Universität, Humboldtstraße 10, 69 Jena, DDR

Prof. Dr. Hartwig Höcker, Deutsches Wollforschungs-Institut e. V. an der Technischen Hochschule Aachen, Veltmanplatz 8, 5100 Aachen, FRG

Prof. Hans-Henning Kausch, Laboratoire de Polymères, Ecole Polytechnique Fédérale de Lausanne, 32. ch. de Bellerive, 1007 Lausanne, Switzerland

Prof. Joseph P. Kennedy, Institute of Polymer Science, The University of Akron. Akron. Ohio 44325, U.S.A.

Prof. Anthony Ledwith, Department of Inorganic, Physical and Industrial Chemistry, University of Liverpool, Liverpool L69 3BX, England

Prof. Seizo Okamura, No. 24, Minamigoshi-Machi Okazaki, Sakyo-Ku, Kyoto 606, Japan

Prof. Salvador Olivé, Chemical Department, University of California, San Diego, La Jolla, CA 92037, U.S.A.

Prof. Charles G. Overberger, Department of Chemistry. The University of Michigan, Ann Arbor, Michigan 48109, U.S.A.

Prof. Helmut Ringsdorf, Institut für Organische Chemie, Johannes-Gutenberg-Universität, J.-J.-Becher Weg 18–20, 6500 Mainz, FRG

Prof. Takeo Saegusa, Department of Synthetic Chemistry, Faculty of Engineering, Kyoto University, Yoshida, Kyoto, Japan

Prof. John L. Schrag, University of Wisconsin, Department of Chemistry, 1101 University Avenue, Madison, Wisconsin 53706, U.S.A.

Prof. William P. Slichter, Executive, Director, Research-Materials Science and Engineering Division AT & T Bell Laboratories, 600 Mountain Avenue, Murray Hill, NJ 07974, U.S.A.

Prof. John K. Stille, Department of Chemistry. Colorado State University, Fort Collins, Colorado 80523, U.S.A.

Table of Contents

Epoxy Molding Compounds as Encapsulation Materials for Microelectronic Devices
N. Kinjo, M. Ogata, K. Nishi, and A. Kaneda 1

Synthesis and Structure of Macromolecular Topological Compounds
Yu. S. Lipatov, T. E. Lipatova, and L. F. Kosyanchuk . . . 49

Reactions and Photodynamics in Polymer Solids
K. Horie and I. Mita 77

Thermotropic Mesophases in Element-Organic Polymers
Yu. K. Godovsky and V. S. Papkov 129

Author Index Volume 1–88 181

Subject Index 193

Epoxy Molding Compounds as Encapsulation Materials for Microelectronic Devices

Noriyuki Kinjo[1], Masatsugu Ogata[1], Kunihiko Nishi[2] and Aizou Kaneda[3]

Epoxy molding compounds for microelectronic devices have been and will continue to be the main stay of encapsulation materials in view of their cost and productivity advantages. On the other hand, as chip sizes become larger due to increased integration of devices, compacter packages are in demand to realize the higher integration. Advances in surface mounting technologies demand encapsulation materials which have extremely low thermal stress and excellent stability at the elevated temperatures used in reflow soldering. Many new technologies have been developed to meet these desires in microelectronic encapsulation. This paper reviews the fundamental properties of epoxy molding compounds and explains the roles and characteristics of the following components which represent encapsulation materials; epoxy resins, hardeners, accelerators, flexibilizers, fillers, flame retardants, coupling agents, and release agents.

1 Introduction 3

2 Epoxy Molding Compounds in Microelectronics Encapsulation 4
   2.1 Structure and Production Processes of Plastic Packages 4
   2.2 Requirements for Encapsulation Materials 6
      2.2.1 Productivity 6
      2.2.2 Reliability 7

3 Raw Materials for Epoxy Molding Compounds 18
   3.1 Epoxy Resins 18
   3.2 Hardeners 19
   3.3 Accelerators 20
   3.4 Flexibilizers 20
   3.5 Fillers 22
   3.6 Flame Retardants 23
   3.7 Other Agents 24

---

[1] Hitachi Research Laboratory, Hitachi Ltd. 4026 Kuji-cho, Hitachi-shi, Ibaraki-ken, 319-12, Japan
[2] Musashi Works, Hitachi Ltd., 1450, Josuihon-cho, Kodaira-shi, Tokyo, 187, Japan
[3] Production Engineering Research Laboratory, Hitachi Ltd. 292 Yoshida-cho, Totsuka-ku, Yokohama, 244, Japan

4 Properties of Base Epoxy Resins 25
  4.1 Catalytic Effects of Accelerators 26
  4.2 Influence of Crosslinking on Physical Properties 29
  4.3 Interpretation of Curing and Thermal Shrinkages 32

5 Development of Highly Reliable Encapsulation Materials 34
  5.1 Lowering Thermal Stress in Molding Compounds 35
  5.2 Moisture-Proof Reliability 37
  5.3 High Heat Resistance 38
  5.4 Lowering α-Ray Sources 39

6 Rheological Characteristics and Mold Filling Dynamics of Epoxy Molding Compounds 39
  6.1 Flow Characteristics of Epoxy-Filler Systems 40
  6.2 Uniform Mold Filling Method 43

7 New Approachs to Encapsulation Materials 44
  7.1 Poly-p-Phenylene Sulfide 44
  7.2 Polyimides 44

8 Summary 45

9 References 46

1 Introduction

Semiconductors, starting with germanium diodes in the 1940s, have developed into transistors, replacing electric tubes, making circuits smaller, and lowering power dissipation. Then, in the 1960's, integrated circuits (ICs) which integrate many diodes, transistors and resistors on one chip led to larger scale integrated circuits known as LSIs and VLSIs. These circuits introduced today's computer age by making circuits much smaller, more functional, and more reliable.

The packages in which these devices are used provide protection from the surroundings and make possible connection of devices with printed circuit boards. The packaging technology for diodes and transistors was can sealing at first. With improvements in passivation technology, plastic encapsulation technology was introduced [1,2]. A short history of packages and devices is shown in Fig. 1. It was plastic packaging technology that made it possible to supply devices more cheaply and in large amounts.

Since higher pin counts and precise exterior dimensions were required for ICs together with higher productivity, a transfer molding process employing molding powder was introduced [3]. Compared with hermetic packages such as cans and ceramics, the plastic package technology has advantages of mass production and

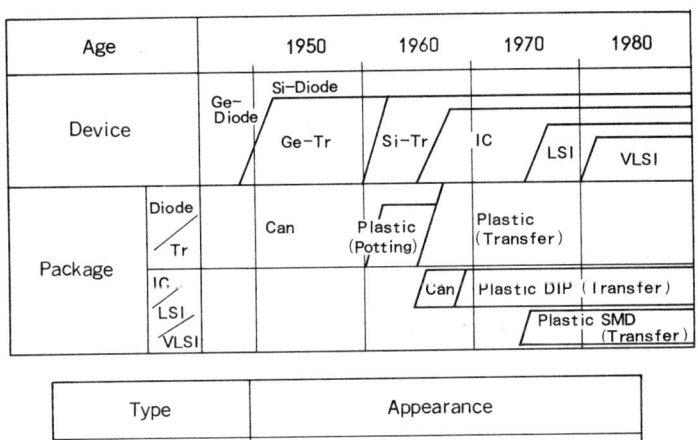

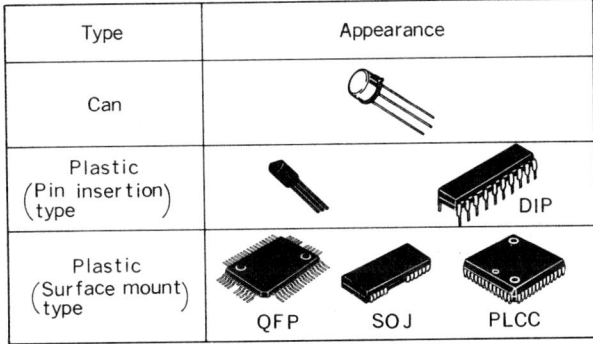

DIP = Dual Inline Package
QFP = Quad Flat Package
PLCC = Plastic Leaded Chip Carrier
SOJ = Small outline J-lead package

Fig. 1. History of semiconductors and their packages

lower cost, even though there are disadvantages in reliability, especially moisture resistance. The latter arises because plastics inherently absorb and permeate moisture more than metals and ceramics do. Then emphasis was placed on development of improved powders to achieve higher moisture resistance while keeping aspects of high purity, low thermal stress, and good adhesion to the chip.

The success of plastic technology in achieving higher reliability and mass production has led to wider applications in consumer products such as electronic calculators, radios, televisions, and video recorders and in industrial products such as personal computers, office automation equipment, and computers. Today, more than 80% of all semiconductor devices are encapsulated by epoxy molding compounds.

The history of the semiconductor development has involved the challenge of achieving higher functionality, while making sizes smaller. The functions implemented on chip have been increasing steadily, which inevitably has increased the pin count and chip size. On the other hand, package size has become smaller and thinner to make compacter mounting on circuit boards. This trend has required low thermal stress encapsulation materials.

In the following, packaging materials, especially recent plastic encapsulation materials, are discussed.

2 Epoxy Molding Compounds in Microelectronics Encapsulation

2.1 Structure and Production Processes of Plastic Packages

The plastic package structure and assembly flow of semiconductor devices are shown in Fig. 2. The semiconductor chip is mounted on a tab lead frame, and the chip bonding pad is connected by a gold wire to the frame. The assembled frame is placed on a molding die, and then transfer molded. After that the outer leads are trimmed, formed, and metal finished (e.g. solder dipping or plating). The final products are tested electrically and then shipped to the buyers.

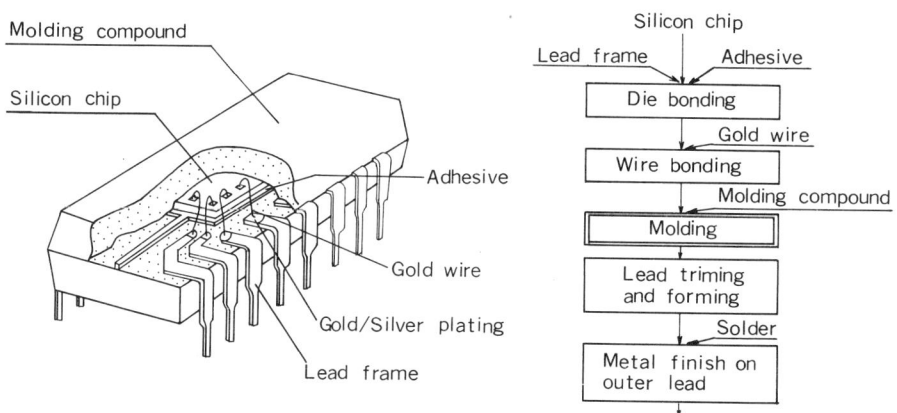

Fig. 2. Plastic package structure and assembly flow of devices

Table 1. Component materials for plastic packages

Material		T.E.C. $\alpha \times 10^6$ (K^{-1})	Young's modulus E (GPa)
Chip	Silicon	3–4	190
	Aluminum	23.0	70
Connector	Gold	14.0	78
Lead frame	Alloy 42	4.5–5.8	154
	Copper	17.0	125
Encapsulation material	Silicone	20–35	10–15
	Epoxy	15–35	10–20

As shown in Table 1, the component materials have different thermal expansion coefficients, and careful choice of materials is necessary for minimizing internal stress of the package, based upon its structure analysis.

For plastic encapsulation materials, thermosetting resin is dominant in semiconductor applications. Thermoplastic polymer generally has a high viscosity which can potentially cause a wire sweep problem. Furthermore, it softens or even melts at higher temperatures. On the other hand, thermosetting resins have a lower melt viscosity before curing and higher heat resistance in the cured state.

Molding compounds consisting of silicone resins were first used as an encapsulation material because of their good adhesion to the chip and lead frame and their higher heat resistance [1, 2]. But about 15 years ago, low cost, high reliability epoxy molding compounds were introduced. Improvements have been made steadily in these compounds as needs have changed and they are now the most common material for plastic packages.

There are various ways to encapsulate such as potting, casting by liquid resin compounds, and transfer molding by powder, with transfer molding being the most popular method to encapsulate ICs or LSIs. The transfer molding process is explained in Fig. 3.

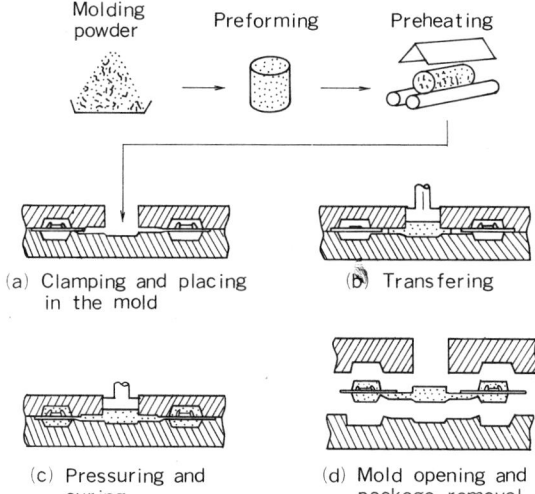

Fig. 3. Transfer molding process

Generally many cavities are arranged along runners in the molding die, and 40 to 200 packages can be molded at one shot. The encapsulation material is first preformed, then preheated by a high frequency apparatus, and put into the pot of the molding die. Lowering a plunger forces the encapsulation material into each cavity through the runners and gates. When curing is completed, the molding die is opened and the packages with attached runners are removed. The molded products are further post-cured.

2.2 Requirements for Encapsulation Materials

Package characteristics such as productivity and reliability reflect the encapsulation material properties desired by the semiconductor manufacturers. Here the relationships between properties of the material and package characteristics are discussed.

2.2.1 Productivity

Important parameters controlling productivity are the cure time, cleaning interval of the molding die, yield (main failure modes being insufficient molding and pin holes), and flash on the lead. It is also important to keep good quality without wire sweep, inner voids, or inner lead deformation.

The curing process subsequent to melting of molding compounds in transfer molding is illustrated in Fig. 4. The letters in parentheses in the following discussion refer to points in the graph. Curing time is generally set so that hardness reaches a certain value (GH). As this curing time is closely related to productivity, faster times are better, but sometimes a trade-off must be made with shelf life (generally the compound is stored under refrigeration). So it is necessary to formulate the material which cures fast and has a longer refrigeration shelf life.

Mold cleaning is generally done after several hundred shots, but before bad mold release or mold stain is observed. Longer intervals between mold cleaning are desirable.

Insufficient molding, inner voids, wire sweep and lead deformation are generally related to flow characteristics and gel time. A material with lower melt viscosity (D), and longer gel time (F) will have fewer failures.

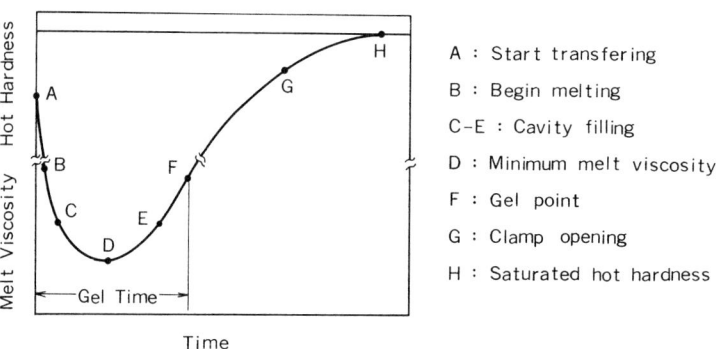

Fig. 4. Schematic illustration of melt and cure properties of encapsulation material in transfer molding

Flash is a very thin plastic film on the lead frame which bleeds out of the package body during molding. As this sometimes causes dewetting of solder plating or solder dipping, it must be removed, so a shorter flash level is required.

2.2.2 Reliability

Reliability of semiconductor devices is determined by two considerations. The first one is early failures, usually caused by fluctuations in production, and the second one is fatigue, related to package design. Accelerated tests are necessary to detect those failures earlier. Many test methods have been proposed for development and assurance on the manufacturer's side [4, 5], and for incoming inspection on the user's side [3, 6, 7]. In this section, correlations between failures in reliability tests and the factors are discussed in detail so as to introduce requirements on encapsulation material from reliability aspects.

1) Temperature Cycling Performance

The temperature cycling test for accelerating ambient conditions is usually done in equipment having two chambers for higher and lower temperature, such as 150 °C and −55 °C for instance. The temperature change in the package, consisting of different materials, causes stress between different materials which is calculated in an elastic model as follows [8].

$$\sigma = \int E(T) \times \alpha(T) \, dT \tag{1}$$

σ: stress (MPa)
α: thermal expansion coefficient (K^{-1})
E: Young's modulus (MPa)
dT: temperature difference (K)

To determine mold stress, three methods have been devised, which are explained below.

a) Double Cylindrical Method

A cross-section of the test apparatus employing a strain gauge to determine encapsulant stress is given in Fig. 5 [9]. The measuring device is a tube with known

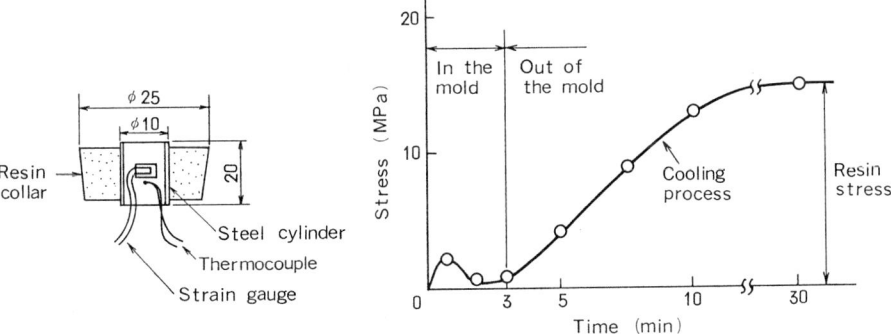

Fig. 5. Stress evaluation by double cylindrical method

thermal expansion coefficient (for instance, steel with T.E.C. of 1.1×10^{-5} K^{-1}) and a strain gauge attached to the inner wall. Before making measurements, the steel tube is calibrated using oil pressure. Then this steel tube is molded by encapsulation material to form a collar. After molding, the collar is cooled to room temperature, and resultant compression is detected as a distortion about the collar circumference. The stress caused by the collared encapsulant changes with decreasing temperature which is schematically shown in Fig. 5. This method has the advantage of being able to evalute stress change during the cooling process.

b) Piezoresistive Method [13]

The piezoresistive effect is a phenomenon occurring when stress is applied to a crystal causing strain inside the crystal and changing the resistivity, according to the equation below [10–12].

$$\delta \varrho_i / \varrho_i = \Sigma \, \pi_{ij} \sigma_j \quad (i, j = 16) \tag{2}$$

ϱ: resistivity component (tensor)
$\delta\varrho$: resistivity change (tensor)
π: piezoresistive factor (tensor)
σ: stress applied to crystal (tensor)

The piezoresistive factor varies significantly, according to the dopant type and crystal axes [14, 15]. In order to simplify the calculation, p-type and n-type piezoresistors are arranged in different directions on the test device as is shown in Fig. 6. The resistance change before and after molding is converted into various stresses such as stress perpendicular to the chip surface, shear stress between the chip and encapsulation material, and compressive stress to the chip. By arraying this kind of stress-sensitive unit at various points on the chip, the stress distribution on the chip surface can be measured. The results contribute to predictions of crack occurence sites of molded devices.

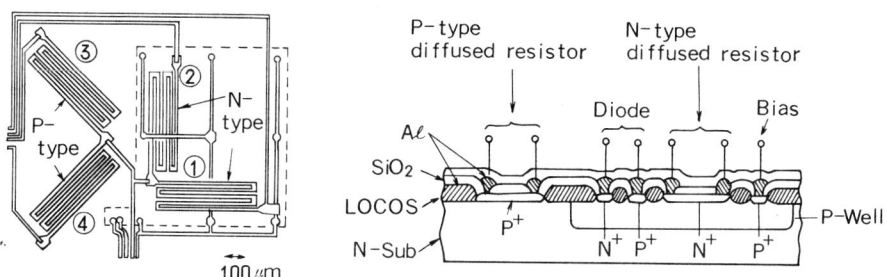

Fig. 6. Piezoresistor test devcice to evaluate the thermal stress of molding compounds

c) Stress Calculation from Physical Properties

Stress is calculated from the observed thermal expansion coefficient, Young's modulus as a function of temperature, and Poisson's ratio [15, 16]. For example, in the case of the double cylindrical model as shown in Fig. 5, the equation is as below [9]

$$\sigma = \int \frac{\alpha_r(T) - \alpha_s(T)}{\frac{1}{E_r(T)}\left(\frac{1+(R_s/R_r)^2}{1-(R_s/R_r)^2} + v_r\right) + \frac{1}{E_s(T)}(1-v_s)} dT \quad (3)$$

σ: stress (MPa)
α: thermal expansion coefficient (K^{-1}).
E: Young's modulus (MPa)
v: Poisson's ratio
R: outer radius (m)
subscripts r, s: resin, and steel respectively

As the molding temperature is higher than the glass transition temperature, stress relaxation occurs during the cooling process. So, by just calculating from the temperature dependence of α and E, the calculated stress is bigger than that observed. Then it is necessary to correct for this by evaluating the stress relaxation modulus. Calculations without and with stress relaxation and observed results are shown in Table 2. The result suggests that stress relaxation actually takes place in the molded devices.

There are advantages and disadvantages in each of the three methods to evaluate the stress occurring at molding. The calculation method is the most popular because it is not necessary to have special equipment, but if stress relaxation is not taken into account, an accurate estimation is impossible. The double cylindrical method provides an accurate stress value including stress relaxation. Although the piezoresistive method is somewhat troublesome regarding device and equipment preparation, it can be used to evaluate actual stress distribution inside a chip directly.

The failure modes in the temperature cycling test are shown in Fig. 7. Failures are of two types: one includes defects in the chip itself caused by resin stress such as a passivation defect [9, 17], aluminum metallization deformation [17, 19], passivation cracks underneath the bonding pad [20], or silicon chip crack [18], and the other includes the defects in the encapsulation material itself, caused by its own stress [21, 22]. The stress applied to the chip inside the package calculated by FEM (Finite Element Method) [25] are also given in Fig. 7. Corrective action against those failures involves

Table 2. Comparison between calculated and measured stresses

Model/Measurement	σ (MPa)
Calculated result without stress relaxation model	20
Calculated result with stress relaxation model	15
Measured by double cylindrical method	12

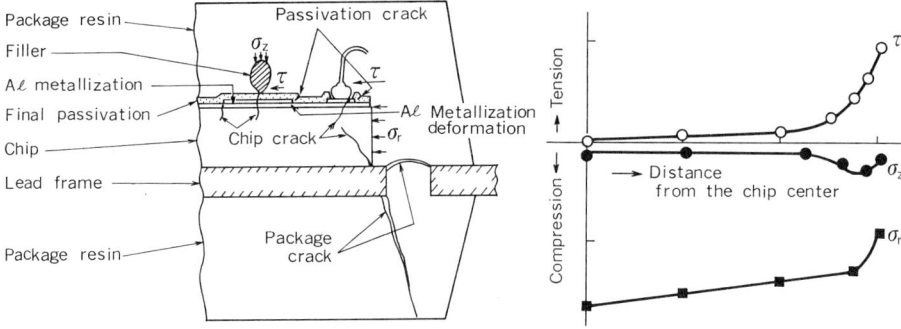

Fig. 7. Failure modes in temperature cycling test and stress distribution of molded chips

lessening the molding stress by decreasing the thermal expansion coefficient and the Young's modulus. Recently, a different type of passivation crack has been found which is caused by stress concentration along the filler edge. For this failure mode, FEM analysis was done regarding the filler particle size and shape which showed that fine particle size and round shape could eliminate the failure [23, 24].

Package cracks occur when a micro-crack propagates from inside to the package surface during the temperature cycling test, due to the thermal stress caused by the difference in thermal expansion coefficients between the encapsulation material and the silicon chip. Furthermore, the appearance of the cross-section of the breakage indicates that the filler is broken together with the thermosetting resin. Considering that the higher the thermal stress becomes in the composite, the lower the temperature is, the crack should propagate at a lower temperature.

It was pointed out previously that a strong correlation was found, also for polymer materials, between the fatigue crack propagation rate da/dN and the stress intensity factor range ΔK prevailing at advancing crack tip [26]:

$$\frac{da}{dN} = A(\Delta K)^n \qquad (4)$$

where A and n are parameters dependent on material variables, mean stress and environment. Figure 8 shows the fatigue crack propagation rate versus stress intensity factor range for an epoxy molding compound [27], satisfying Eq. (4). The n-value obtained from Fig. 8 is about 20 which is much larger than those of linear polymers ranging from 2 to 8. This means that crack propagation rate of epoxy molding compounds is affected by advancing crack tip more easily than usual linear polymers are. Reduction of thermal stresses and making base epoxy resins tough, which will be discussed later, are countermeasures against crack propagation.

2) Moisture Resistance

Since absorption and permeation of water through polymer materials are inevitable, the most serious problem in plastic encapsulated devices is aluminum metallization corrosion in moist atmospheres, which leads to functional failures of the devices [29, 30].

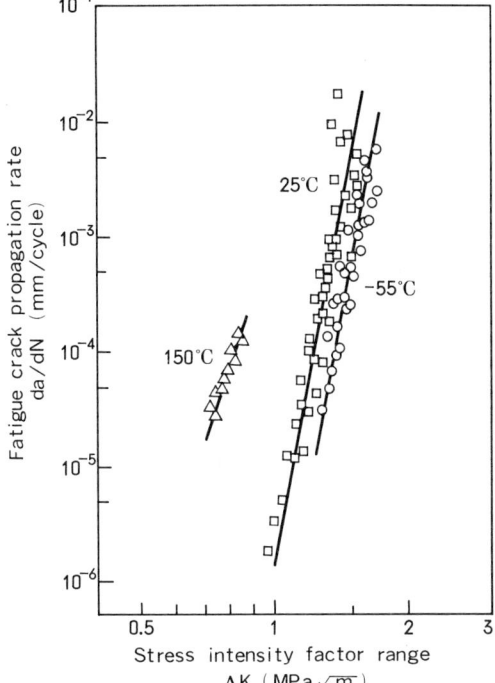

Fig. 8. Fatigue crack propagation rate of encapsulants

To detect aluminum corrosion, various accelerated tests are available, typical ones being;
1) PCT: Pressure Cooker Test done at 121 °C, 2 atm steam;
2) HHBT: High temperature High humidity Bias test done for example at 85 °C, 85% RH bias;
3) HHT: High temperature High Humidity Test done for example at 65 °C, 95% RH.

Cumulative failures are shown on a Weibull distribution plot in general, and the factor m is obtained. When m > 1, the failure mode is fatigue owing to design, and when m < 1, the failure mode is early failure caused by process fluctuation. Generally, m is bigger than 1 for PCT, which means the test detects fatigue occurrence. However, HHBT and HHT make m smaller than 1, which means these tests detect early failures.

There are two paths for moisture penetration. One is through the material itself and the other one is through the interface of the encapsulation material and lead frame, gold wire and the chip. Moisture penetration through the bulk is calculated by Fick's law using the weight gain of a disc stored under high temperature, high humidity conditions [35]. Fick's law is

$$\frac{C(t)}{C_0} = 1 - \frac{4}{\pi} \exp\left(\frac{-\pi^2 Dt}{l^2}\right) \tag{5}$$

C: moisture concentration
D: diffusion constant of moisture

After getting D, the moisture concentration on the surface of the chip is calculated by Eq. (5). To detect moisture penetration to the chip surface, leakage current measurements on a test device with parallel metallization are convenient [36, 37]. The current differs depending on the distance between the metallization lines and on the place on the chip surface. When especially large leakage currents are observed, aluminum corrosion has probably occurred [33, 34]. Unusually large leakage currents are observed mainly on the periphery of the chip surface. Since this occurrence is rather similar to that of shear stress on the chip as shown in Fig. 7, it suggests that delamination (i.e. poor adhesion between the chip and encapsulation material) is caused by shear stress.

Figure 9 shows the relationship between chip size and aluminum corrosion rate [9]. As chip size increases, rate increases rapidly for compound A which is not a low thermal stress molding compound. On the other hand, rate does not increase so much for compound B which is such a compound. As the chip surface is overcoated by passivation film such as SiO_2 and PSG (phosphorus silicate glass) to protect against contamination and damage such as scar and cracks, this passivation film sometimes cracks and its apriori defects are developed by the molding stress. Such problems are also found in the periphery of the chip surface.

These findings all point out that both leakage current and passivation defects are mainly caused by shear stress, and moisture penetrating into the chip surface reaches the aluminum metallization layer through passivation defects, eventually resulting aluminum corrosion.

The failure mode of aluminum corrosion is of two types. One is anodic corrosion, sometimes called pitting, and the other one is cathodic corrosion.

a) Anodic Corrosion [31]

It is well known that aluminum corrodes in both acid and alkaline solutions, but not in neutral ones (pH = 4 to 8) because of aluminum oxide formation which serves as a passivation film [32]. Ideally the pH of the water extracted solution of encapsulation material is controlled in this region. But when halide ions like Cl^- and Br^- exist in

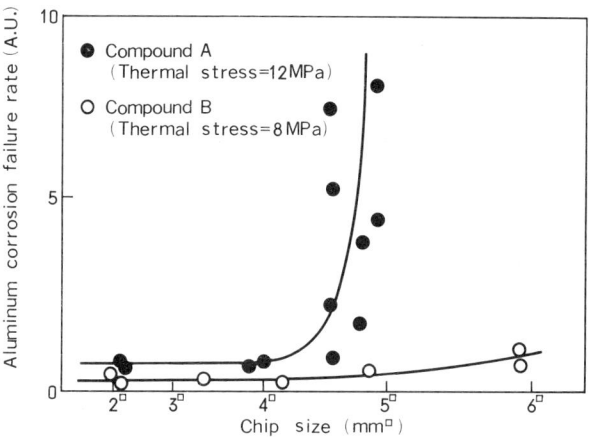

Fig. 9. Aluminum corrosion vs. chip size after 1000 h at 65 °C, 95% RH

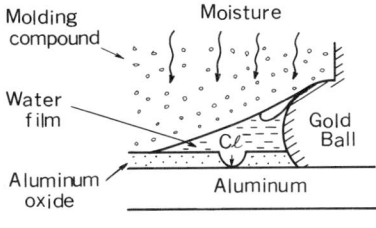

(a) Water film is formed by penetrating moisture.

(b) Cl ion in the water film attacks aluminum oxide to form pits.

(c) Local anodic reaction:
$$Al \longrightarrow Al^{3+} + 3e^-$$

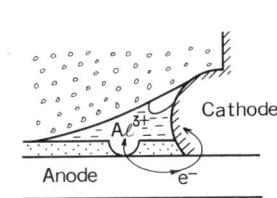

(d) Local cathodic reaction:
$$\frac{1}{2}O_2 + H_2O + 2e^- \longrightarrow 2OH^-$$
$$M^+ + e^- \longrightarrow M \quad (M^+: \text{Metal ion})$$

(e) $Al^{3+} + 3(OH)^- \longrightarrow Al(OH)_3$

Fig. 10. Aluminum corrosion mechanism

large amounts, they attack the aluminum oxide, allowing corrosion to occur. This has been confirmed by chloride ion detection in the corroded area of aluminum by XMA. A schematic illustration of this is in Fig. 10. Investigation of the effect of chloride in the encapsulation material on aluminum corrosion showed a large effect [38, 39]. The chloride source is mainly the base epoxy resin, e.g. as is synthesized from epichlorhidoline. Early epoxies contained about 1000 ppm of hydrolyzable chloride, but more recent ones contain only 100 ppm. Then countermeasures for the chloride effect must look at external means, rather than reducing further the chloride present in the epoxies.

In summary for anodic corrosion, water molecules permeate through the encapsulation material to reach the chip surface. Delamination between the lead frame and encapsulation material must accelerate moisture penetration into devices. Delamination between chips and encapsulation material allows water film formation. If halides, such as chloride ion are present in the water film, the corrosion of aluminum is accelerated. Corrosion takes place mainly on aluminum of the bonding pad, since the chip is usually coated with passivation film. But when passivation defects exist, moisture penetrates through them and aluminum corrosion of the wiring conductor occurs. Then improvements in adhesion, and lowering molding stress and chloride content are important.

b) Cathodic Corrosion [39–42]

When stored at high temperature and high humidity under a bias, a small leakage current flows between the metallization layers, through the encapsulation material. As a result, the following reactions occur on the negatively biased surface of the grain boundary.

$$H_2O + e^- \rightarrow OH^- + \frac{1}{2}H_2$$

$$OH^- + Al + H_2O \rightarrow (AlO_2)^- + \frac{3}{2}H_2$$

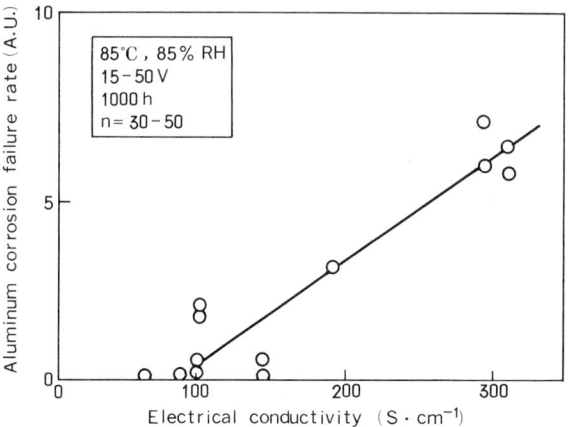

Fig. 11. Cathodic corrosion rate of aluminum increases with the contents of ionic extracts from molding compounds into water

The $(AlO_2)^-$ and the hydrogen gas generated cause disconnection of the metallization layers [43, 44].

The failure due to the cathodic corrosion of aluminum is also accelerated by ionic impurities existing in the water film on the bonding pad area. Ion chromatographic analysis showed that the extracts from molding compound into water were hydrolyzed products such as $R-COO^-$ and inorganic ions from polymer. Since leakage current between the biased metallization layers is subject to the conductivity of water films, cathodic corrosion should be closely related to the conductivity of water films including extracts from molding compound. The experimental results is shown in Fig. 11, where the rate of device failure due to the cathodic corrosion is plotted against the conductivity of water suspended with crashed molding compounds. The higher the conductivity, the greater the probability of failure becomes. This result also suggests that decreasing ionic materials in molding compounds is prerequisite to realizing highly reliable and high quality molding compounds.

3) Electrical Properties

To maintain insulation between pins, volume resistivity must be more than 10^{12} ohm-cm. On the other hand, when devices are operated at higher temperature, leakage occurs depending upon the device and/or encapsulation material. This phenomenon is well known as "parasitic MOS" [45]. Two independent diffusion layers are connected by the channel of a depletion layer which is generated by the induced surface charge as is shown in Fig. 12. Channel formation time has a fairly good correlation with volume resistivity of encapsulation material at higher temperature. So higher volume resistivity at higher temperature is necessary to provide reliable devices.

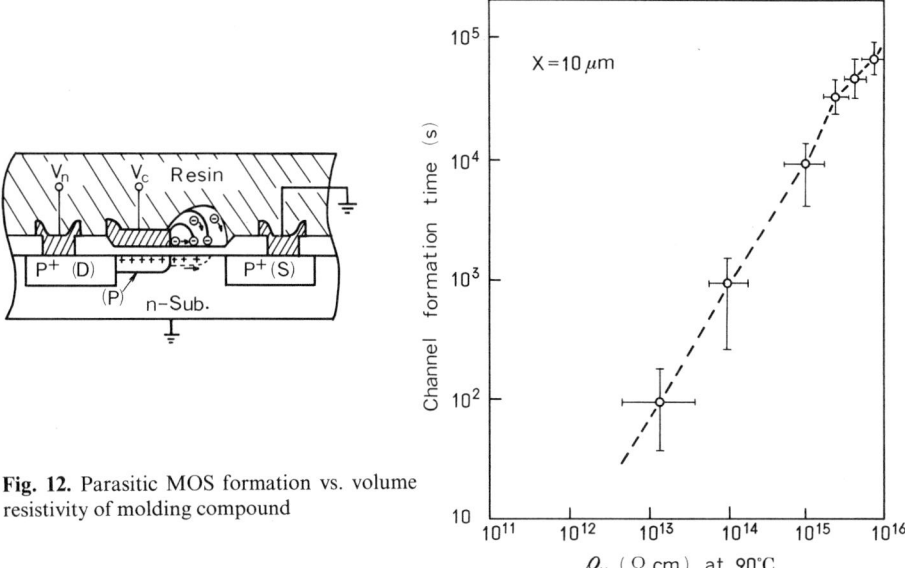

Fig. 12. Parasitic MOS formation vs. volume resistivity of molding compound

4) Thermal Properties

Since semiconductors generate heat when operated, it is necessary to dissipate heat from the package for stable operation. The heat resistance of the package is defined by the equation below.

$$\theta_j - a = \frac{T_j - T_a}{P_d} \qquad (6)$$

$\theta_j - a$: heat resistance (KJ^{-1})
T_j: junction temperature (K)
T_a: ambient temperature (K)
P_d: power dissipation (J)

The V-f method is used to determine heat resistance, and its procedure is described below.
1) A correlation curve between the temperature and forward current of the device is obtained beforehand.
2) The foward current is measured when a certain power is applied.
3) Tj is obtained by making use of the correlation curve and measured forward current.
4) θj — a is calculated from Tj and Pd by Eq. (6).

There are many heat paths from the heat source (device junction), and the heat resistance is their summing up; that is,

$$\theta_j - a = \sum_i \frac{A_i}{\lambda_i \cdot L_i} \qquad (7)$$

A_i: cross-section of heat path (cm^2)
L_i: length of heat path (cm)
λ_i: thermal conductivity of each path (J cm^{-1} s^{-1} K^{-1})

A fairly good correlation is obtained between heat resistance and thermal conductivity of encapsulation material. It shows that a higher thermal conductivity gives a lower heat resistance.

Recently, a new failure mode has been found in the accelerated testing of high temperature storage [47,48]. It shows an increase in the resistance of the bonding pad, or in severe cases disconnection, when a device is stored at high temperatures around 200 °C. It is due to the formation of an aluminum-gold intermetallic compound [118]. Its formation seems to be accelerated by free bromine or chlorine [99-100]. It may be brominated epoxy resin, which is added to maintain flame retardancy at high temperatures, that generates free bromine. To prevent this, it is necessary to formulate epoxy resins which does not decompose at high temperatures.

5) Soft Error Problem

It was first reported in 1978 that memorized information disappeared because of α-rays emitted from the package [49]. Since then various approaches have been taken from both the device side and the package side to reduce this phenomenon which is known as "soft error" [50,51,121,123]. It has the following features.
1) The origin of α-rays is U and/or Th in the silica of the encapsulation material.
2) The energy of the α-rays is from 5 to 9 MeV.
3) When an α-ray of 5 MeV crosses a memory cell, 1.4×10^6 pairs of electrons and holes are generated.
4) In the case of an N-channel MOS, holes flow into the substrate, and electrons flow into the active area. Then neutralization occurs in the memory cell which destroys the memorized information.
5) If information is rewritten, the cell operates normally.

A corrective action is simply to reduce U and Th content in the encapsulation material. The U content level should be less than 1 ppb, for example around 0.5 ppb for a 1 Mbit D-RAM.

6) Soldering Problems on Surface Mount Type Devices

As use of surface mount type devices increases, replacing pin insertion types soldering methods are changing from wave soldering to reflow soldering methods such as vapor phase soldering and IR soldering. Although reflow soldering has advantages of easy mass production and reliable soldering, heat is sometimes applied to the plastic body, unlike in wave soldering. When a package absorbs moisture, and is then put into a soldering bath, the rapid expansion of the moisture causes delamination of the interface which results in lower moisture resistance, and sometimes even cause package cracking [52,53]. The temperature dependence of the strength of the encapsulation material is given in Fig. 13, along with calculated stress at various moisture contents inside the package. At around 215 °C which is a common temperature for vapour phase soldering (VPS), the stress exceeds the strength of the encapsulation, causing package cracking. This package cracking occurs mostly on the tab side not on the chip side. From these results, it is necessary to increase the material strength and adhesion to the metal, while decreasing moisture peneration. It has been also confirmed that moisture resistance performance after soldering can be improved by using a low thermal stress encapsulation materials. Scanning acoustic tomography (SAT) is

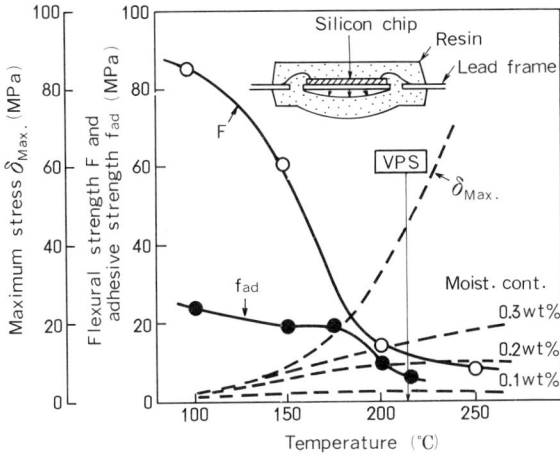

Fig. 13. Stress calculation and strength of resin

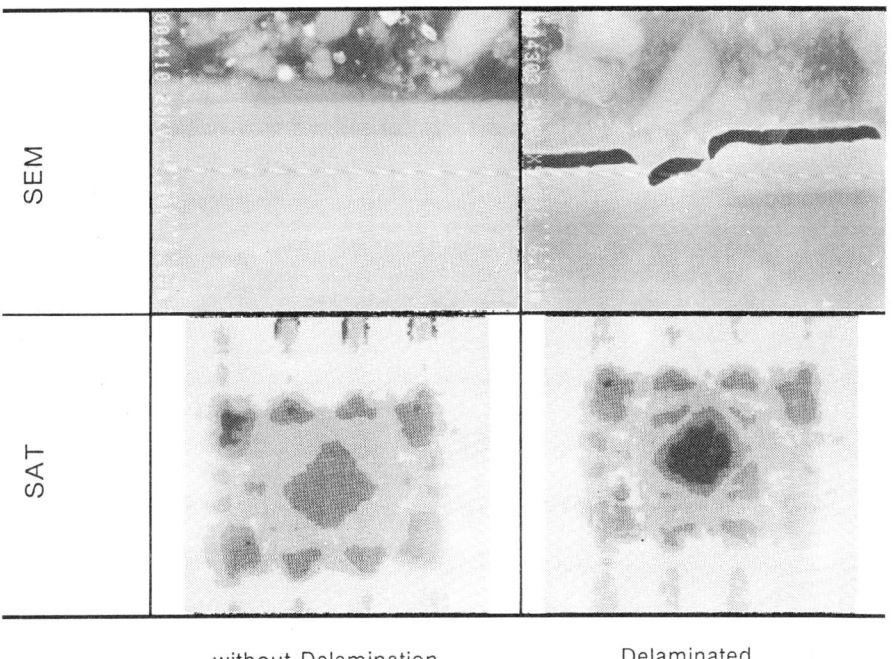

Fig. 14. Inspection of the delamination between Si chip and encapsulant by SEM and SAT (Scanning Acoustic Tomography) [53]

very helpful non-destructive method for observing the delamination area inside the molded devices. Example of SAT studies are shown in Fig. 14, comparing with the results of SEM observations.

3 Raw Materials for Epoxy Molding Compounds

Epoxy molding compounds in microelectronics encapsulation are usually composed of more than ten kinds of raw materials [54-56], each of which has its own special functions. These raw materials and their functions are discussed in this section.

3.1 Epoxy Resins

Fundamental characteristics of the molding compound in microelectronics encapsulation to consider are moldability, mechanical and electrical properties, and humidity and heat resistance. These depend significantly on the corresponding characteristics of the base epoxy resins used. Epoxy resins can be classified roughly into two types: structoterminal and structopendant types. Each type includes a variety of epoxy resins having different chemical structures [57]. In these epoxy resins, crosslinking takes place when hardners such anhydrides, amines, phenols or bases are added as curing agents. Since cured resins show very different properties according to the ratio and species of base epoxy resins and hardners [57-62], various combinations of base resins and hardners are tailored to fit the molding methods used or the LSI device requirements.

Table 3. Characteristics of typical epoxides, hardeners and accelerators

Material		General characteristics	
		Before curing	After curing
Epoxide	Bisphenol A types	Range from liquid to solid	Low Tg, excellent flexibility
	Phenolnovolac types	Solid excellent reactivity	High Tg, excellent heat resistance
	o-Cresolnovolac types	Solid	High Tg, excellent heat and humidity resistance
	Cycloaliphatic types	Liquid (low viscosity)	High Tg, excellent electrical properties and weathering resistance
Hardener	Amine compounds	Excellent reactivity poor storage stability	Poor electrical properties
	Acid anhydrides	Poor storage stability (easily hydrolyzed)	Excellent electrical properties at elevated temperature
	Phenolic compounds	Excellent moldability	Excellent electrical properties, heat and humidity resistance
Accelerator	Amine compounds	Excellent acceleration poor storage stability	Poor electrical properties and humidity resistance
	Organophosphine compounds	Excellent acceleration mediocre storage stability	Excellent electrical properties, and heat and humidity resistance
	Lewis base salts	Mediocre acceleration excellent storage stability	Excellent electrical properties, and heat and humidity resistance

General properties of typical epoxy resins are summarized in Table 3. Molding methods for LSI devices currently practised are dipping, casting, potting, and low pressure transfer molding [1-3, 55, 63]. Except for the last one, the molding is done at comparatively low temperatures near room temperature, so base resins with low viscosity and high fluidity are required. Therefore, liquid bisphenol A type epoxy resins or alicyclic epoxy resins as base materials, and liquid amines or acid anhydrides as hardners are usually used. In the case of the low pressure transfer method in which heated molding materials are mechanically transferred into a mold die under pressure, solid types are preferred for easier handling of both base materials and hardners; examples include solid bisphenol A type epoxy resins or novolac epoxy resins, and solid amines, acid anhydrides or phenol compounds [3, 64]. Recently, however, as o-cresolnovolac epoxy resins have advantageous pot lifes, formability, and humidity and heat resistance, they are becoming most commonly used as base materials when combined with phenolnovolac resin hardners [54]. In addition, flame retardancy is also required for molding materials as mentioned previously. Details of flame retardancy are discussed later, it suffices here to note only that some of the base materials employ brominated bisphenol-A-epoxy or brominated phenolnovolac epoxy resins as flame retardant agents [97-100].

3.2 Hardeners

Curing reactions of epoxy resins are considered below by the type of hardeners used [57, 65-67]. As examples, crosslinking through the epoxy groups can be achieved through ring-opening polymerization when using amines as hardeners, while crosslinking occurs primarily through the hydroxyl groups when acid anhydrides are used.

For amine types hardeners:

$$R_1-CH-CH_2 + H_2N-R_2 \longrightarrow R_1-CH-CH_2-NH-R_2 \longrightarrow R_1-CH-CH_2-N(R_2)-CH_2-CH-R_1$$
$$\overset{\diagdown O \diagup}{} \qquad \overset{|}{OH} \qquad \overset{|}{OH} \qquad \overset{|}{OH}$$

For acid anhydride types:

For phenol types:

$$R_1-CH-CH_2 + HO-\langle \text{benzene} \rangle-R_2 \rightarrow R_1-CH-CH_2-O-\langle \text{benzene} \rangle-R_2 \rightarrow$$
$$\underset{O}{\diagdown\diagup} \qquad \qquad \qquad \underset{OH}{|}$$

(with further reaction giving the di-substituted product shown)

Curing reactions of alicyclic polyamines with epoxy resins occur easily enough to allow carrying out at room temperatures. However, aromatic polyamines, acid anhydrides and phenol compounds usually need hardening accelerators, since their curing reactions require long times, even at elevated temperatures. Table 3 summarizes the general characteristics of these hardeners.

3.3 Accelerators

Accelerators usually used are amines, imidazoles, organophosphines, urea derivatives, or Lewis bases and their organic salts [68-75]. Usual accelerators, however, tend to deteriorate pot life or moldability of molding materials because their reactions are initiated at comparatively low temperatures. Thus, better hardening accelerators have a latent characteristic that promotes rapid curing of resins when heated to a specific temperature, but at lower temperatures they do not [76-79]. Many types of such latent accelerators have been developed [72-75]. Table 3 compares characteristics of some molding materials using such latent accelerators. Proper selection of hardening accelerators decides the reliability of molding materials, because of their significant influence on hardening ability and other properties, in addition to pot life [68, 76-78].

3.4 Flexibilizers

Molding materials with a large quantity of inorganic fillers included in their base epoxy resins are rigid and breakable in the cured state. Thermal expansion coefficients of their cured resins are far greater than those of silicon chips or the lead frame wiring that make up semiconductors. Therefore, residual thermal stress is contained in the molded package [80-82]. Further heat cycles easily give rise to new thermal stress, causing passivation cracks, variation in device properties, chip damage, package cracks, etc. [9, 17-20]. Such practical problems due to the thermal stress are described in detail in Sect. 5.1 from the viewpoint of solving them.

Flexibilizers are added in order to provide molding materials which are lower in elasticity, greater toughness and flexibility [28, 85, 86]. There are unreactive and reactive type flexibilizers for epoxy resins [57, 87-92]. The former contain phthalic acid ester or chlorinated biphenyl. As for the latter, there are many kinds: mono-epoxides or phenols of which have long alkyl chains, long-chained diamines, dimer acids, and trimer acids, polyols, epoxides derived from alicyclic acids, polyesters, polyamide resins, urethane elastomers, polybutadiene rubbers, silicone rubbers, etc. If these flexibilizers are completely miscible with matrix epoxy resins, transparent one-phase-resins in the cured state are obtained, resulting in a lower Tg and deteriorations in electrical and mechanical properties at elevated temperatures or humidity resistance. Then, often-practised methods for lowering elasticity without lowering Tg are dispersing particulated flexibilizers such as silicone compounds or polybutadiene rubbers in epoxy resins, or forming microphase-separated structures [54, 85, 86]. The final cured products are opaque, hard resins in these cases. The dispersion of rubber particles into matrix polymer phases also improves the toughness of a hard and brittle polymer matrix. SEM microphotographs of fracture surfaces for flexibilized epoxy resins are shown in Fig. 15. Comparison is made for roughness and the structure of resins using both miscible and immiscible type flexibilizers. Figure 16 shows schematically dynamic viscoelastic spectra of flexibilized resins in the cured state, of which one has a one phase structure and the other has a microphase-separated structure. In the systems where immiscible type flexibilizers are used, the elastic modulus in the glassy state

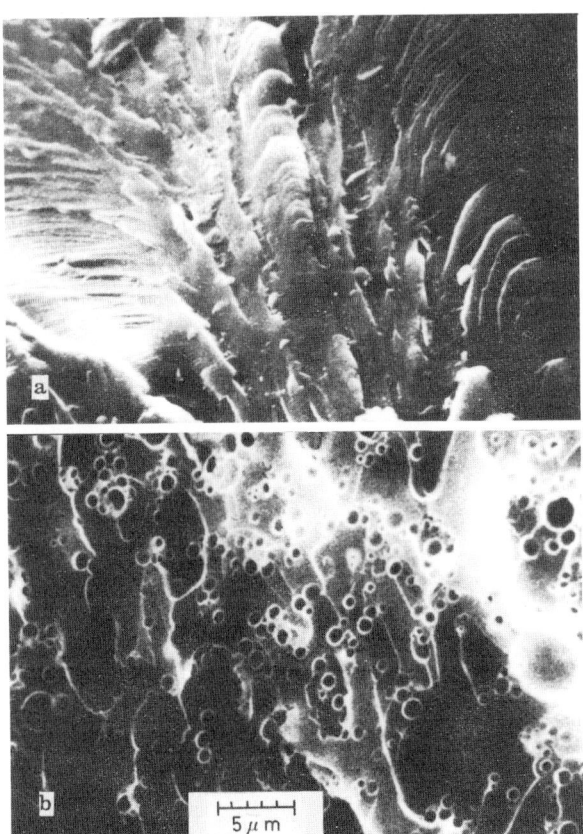

Fig. 15a and b. Fracture surfaces of cured epoxy resins with added flexibilizers

of the resin decreases, while glass transition temperature Tg, that is, the temperature for the maximum loss modulus or tan δ, does not.

It has been found that addition of some types of flexibilizers decreases both elastic moduli and thermal expansion coefficients of molded products, lowering thermal stress significantly [28, 85, 86]. Recent studies have shown that these improvements are brought about by the partial miscible nature of the components. However, such flexibilizers when added to molding materials sometimes bleed to the surface, smearing the appearance of the encapsulated devices and/or dies. Ways to prevent this are being examined, they include chemical modification of flexibilizers with a proper quantity of functional groups which chemically bond with base resins, by using stabilizers to provide finer dispersion of flexibilizer grains, or by using solid rubber particles as flexibilizers.

3.5 Fillers

Fillers are used for lowering shrinkage on curing, decreasing thermal expansion coefficients, improving thermal conductivity, or meeting specific mechanical requirements [55, 57, 94–96]. Although there are many different inorganic fillers that can meet such requirements, ideal fillers must also be chemically stable, and must give good electrical and mechanical properties, and humidity and heat resistance to the cured epoxy molding compound. Crystalline silica powders or quartz (fused silica) powders are most widely used as fillers now [54, 56, 121]. Fig. 17 shows thermal expansion coefficients and thermal conductivity of cured epoxy molding compounds combined with three fillers. Fused silica powder, as is clear from the figure, is advantageous in view of thermal stress, since small thermal expansion coefficients are shown for cured mold-

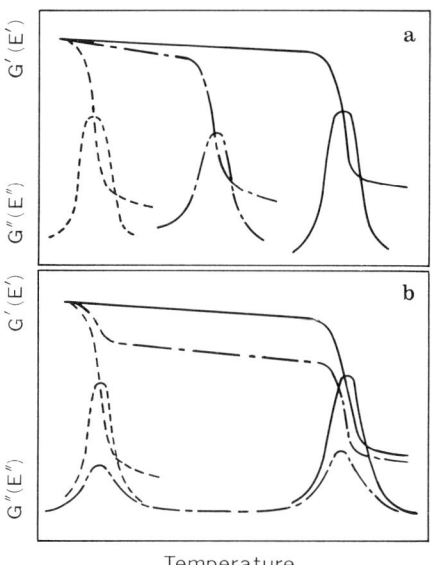

Fig. 16a and b. Schematic representation of the effects of flexibilizers on the viscoelastic behavior of thermosetting resins; ——— hard matrix resin having high Tg; ·········· cured flexibilizer; —·— hard resin and flexibilizer, mixed then cured; a flexibilizers are compatible with the hard matrix resin. Their addition results in the monotonous reduction of the matrix phase Tg; b microphase separation takes place, which results in a peculiar physical structure in which soft rubber particles are scattered throughout a hard matrix phase

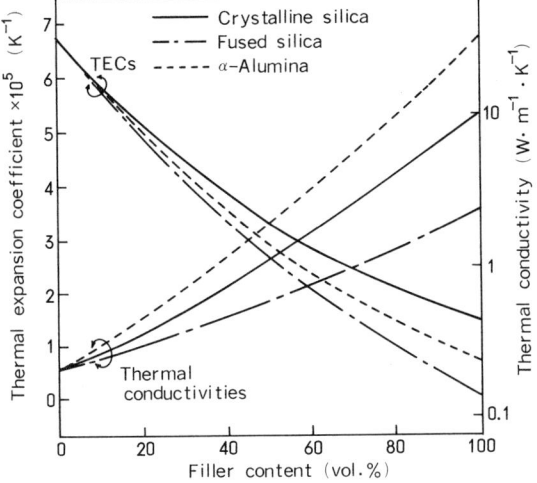

Fig. 17. Thermal expansion coefficients (TECs) and thermal conductivities of cured epoxy resins using various fillers

ing compounds using it as filler. However, in terms of heat dissipation, it is disadvantageous due to its small thermal conductivity. On the other hand, crystalline silica powder has a good heat dissipation characteristic because of its large thermal conductivity, though its compounds' thermal expansion coefficients are a little large. Therefore, fused silica powder is applied to the epoxy molding compounds for LSIs which need thin-skinned encapsulation, while crystalline silica is used for power electronics devices which need considerable heat dissipation.

Increasing the quantity of fillers may contribute to lowering thermal expansion coefficients and increasing thermal conductivity of cured epoxy molding compounds, but too large an increase causes an increase in viscosity and decrease in fluidity of resin compounds before curing [54, 93]. These are serious problems in the actual molding process.

3.6 Flame Retardants

Flame retardancy is required for epoxy molding compounds to insure safety. As a means to give flame retardancy to plastcics, three methods are practised.
1) Improving thermal stability of polymers.
2) Inserting some elements having flame retardancy into polymer chains.
3) Physically mixing flame retardant agents with polymers.

To provide flame retardancy to epoxy molding compounds, methods 1) and 2) are used, where a reactive type agent is employed in the former, and an additive type serves in the latter [97]. Generally, cured resin compounds combined with flame retardant agents tend to show decreased heat resistance, electric properties, and humidity resistance [98, 99]. Therefore, combined uses of reactive type brominated epoxy resins which less deteriorate in these properties, and additive type antimony trioxide are widely used [100–103]. Relationships between quantities of Br and Sb_2O_3 contained, and flame retardancy of cured resins in compliance with UL-94 standards, are shown in Fig. 18. Because

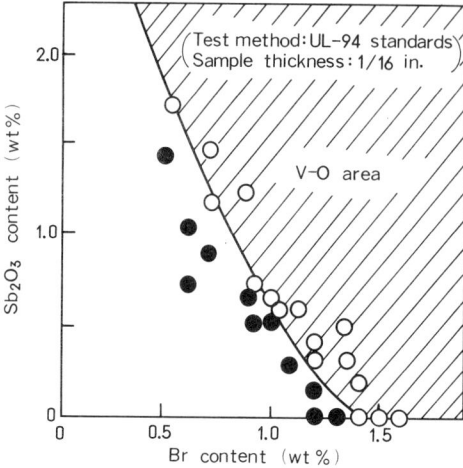

Fig. 18. Relationship between flame retardant content and flame retardancy of cured resins (○ denotes flame retardancy "V-0" and ● denotes "V-1 or HB" tested by UL-94 standard methods)

V-0 requirements can be met for a wide range of Br and Sb_2O_3 amounts, their optimum combination can be determined to balance specific or overall requirements for molding materials.

3.7 Other Agents

In addition to the above-mentioned agents, molding materials often contain other agents, for example, coupling agents to improve affinity or chemical bondage of base epoxy resins with fillers and adhesion to chips or lead frames; release agents to allow easy removal of cured resin compounds from molds; or coloring agents to improve appearance [56]. Careful considerations are also due on these agents because they have a noticeable effect on the properties and reliability of cured resin compounds.

1) Coupling Agents

Although various compounds are known as coupling agents, such as silane [104, 105], titanate [106], aluminum chelates [107], and zircoaluminate [108], the most popular ones are of silane compounds. Not only do these coupling agents of silane offer strong adhesion between base resins and fillers and improved humidity and heat resistance of cured resins, they also contribute to the improvement of adhesion of molding compounds to chips or lead frames. The coupling agent reacts with silanol groups on the fillter surface to form the following [105] structures:

$$\text{Filler surface} \mid Si-O-Si{-}(CH_2{-})_n{-}O-CH_2-CH\underset{O}{-}CH_2$$

Usually, coupling agents are mixed together with base resins and fillers by using rolls or extruders. By giving proper consideration to the mixing process, however, further improvements may be attained by maximizing the coupling agents' effects.

2) Release Agents

While epoxy molding compounds must have good adherence to chips or lead frames on one hand, they must not stick to roll surfaces or extruder walls, or the mold dies. Moreover, shortening the cleaning time at molding is desired from the viewpoint of productivity. To meet such requirements, release agents are added to molding compounds. Organic salts and their metal salts, wax, silicone compounds, and fluorine compounds are generally employed as release agents, but wax is mainly used for epoxy molding compounds [57, 109]. Their release effect depends on fusing temperatures, and sufficient effects are only exhibited at temperatures above fusing. Generally, two kinds of release agents having different fusing temperatures, one lower than 100 °C and the other lower than 180 °C, are used together, because the mixing temperature of the various components to fabricate the molding compound is about 100 °C, while transfer molding of devices is usually carried out at around 180 °C. In order to satisfy the two contradictory properties for molding compounds of good adherence to chips or lead frames, and improved release effects from molds on the other hand, sometimes additional release agents are added to the crushed molding compounds after mixing the various components.

Release agents, however, inevitably have some effect causing deterioration of adhesion strength required for molding compounds in the cured state, damaging the appearance of cured resin compounds, and staining the mold die surface. In order to avoid such problems, careful selection of agents, or their amounts, and handling must be made.

3) Coloring Agents

Inorganic and organic pigments are used for coloring agents. For brightness, organic pigments are better, but for robustness and decolorization, inorganic pigments are better. In many cases, however, inorganic pigments are also preferred because of their better electrical properties, and humidity and heat resistance. As black is preferred for molding compounds, most of the pigments use carbon black [57, 110]. Carbon black is obtained by incomplete combustion or thermal decomposition of hydrocarbons, so it often contains impurities such as sulphur oxides or nitrogen oxides. In addition, carbon black is electrically conductive. Therefore, care should be taken in the purity and quantity of carbon black to be added, because it can influence greatly the curing process, electrical properties, and humidity resistance of the cured resin compounds.

4 Properties of Base Epoxy Resins

Basic resins of today's molding compounds in microelectronics encapsulation are o-cresolnovolac epoxides and brominated DGBA (diglycidyl ether of bisphenol A) cured with phenol-formaldehyde novolac as a hardener (Table 4) [3, 54, 56, 64]. The

Table 4. Formulation of base epoxy resin for recent molding compounds in microelectronics encapsulation

	Name	Parts
Epoxide	Polyglycidyl ether of o-cresol-formaldehyde novolac	90
	Brominated diglycidyl ether of bisphenol A	10
Hardener	Phenol-formaldehyde novolac resin	52
Accelerator	Table 5	0.4 ~ 3.0*

* 3 ~ 12.5 mmols

physical properties of cured epoxies and the characteristics of molding compounds composed of these epoxies differ considerably, depending not only on the species of epoxies and/or hardners, but also on the accelerator kinds [76-79]. Accordingly, it can be said that selection of accelerators provides the key to development of suitable epoxy molding compounds for microelectronics encapsulation, since accelerators have the dominant effects regarding moldability, pot life and shelf life of the molding compounds [68]. This section describes influences of accelerators on some properties of base epoxy resins [111]. Many new developments in accelerators during the last few years have allowed tremendous improvements in microelectronics encapsulation. However, the exact effects of various accelerators in epoxy resins are still not fully understood.

4.1 Catalytic Effects of Accelerators

Many kinds of accelerators have been examined for application to molding compounds in microelectronics encapsulation from the viewpoint of latent catalytic effects. Some typical ones are shown in Table 5; Lewis bases such as imidazoles and Lewis base salts

Table 5. Curability of epoxy resins with various added accelerators

Sample No.	Accelerators	Added amount (mmol)	Gel time (s) at 180 °C	Activation energy (kJ · mol^{-1})
1	N-Benzyldimethylamine (BDMA)	9.0	32	71.2
2	2-Ethyl-4-methylimidazol (EMI)	3.5	33	66.1
3	1.8-Diazabicyclo(5.4.0)-7-undecene (DBU)	7.5	32	66.6
4	N-Methylpiperazine (MP)	5.0	34	74.1
5	Tetramethylbuthylguanidine (TMBG)	12.5	30	72.8
6	Triphenylphosphine (TPP)	3.0	32	67.8
7	Triethylammonium tetraphenylborate (TEA-K)	8.0	32	73.7
8	2-Ethyl-4-methylimidazolium tetraphenylborate (EMI-K)	5.0	35	72.4
9	1.8-Diazabicyclo(5.4.0)-7-undecenium tetraphenylborate (DBU-K)	4.5	35	71.6
10	Tetraphenylphosponium tetraphenylborate (TPP-K)	3.0	34	75.8

such as quarternary phosphonium salts are rather popular. Gel times at 180 °C for epoxy resins mixed with various accelerators are shown in Fig. 19 to compare the catalytic effect for epoxy-phenolic reaction. The acceleration differs with the accelerator type, e.g. EMI, TPP, TPP-K are very effective in giving a shorter gel time even at the comparably low levels. The last column in Table 5 shows the activation energies obtained from Arrhenius plots of gel times in the temperature range from 150 to 180 °C as shown in Fig. 19, when the added amount of each accelerator is controlled so that the gel time at 180 °C is 33 ± 2 s. The results indicate that MP, TEA-K and TPP-K have rather large activation energies, suggesting that their catalytic effects are latent and therefore desirable from a practical viewpoint, as latent catalysts meet the contradictory requirements of long shelf life and fast curing at molding.

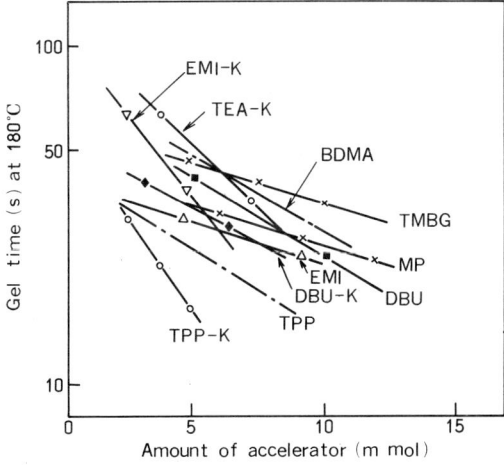

Fig. 19. Relationship between gel time and amount of accelerator

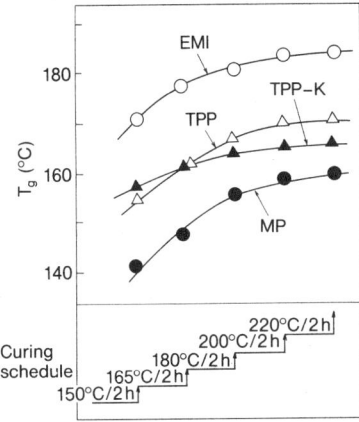

Fig. 20. Effect of curing schedule on Tg

Post-curing at elevated temperatures is usually necessary to accomplish the epoxy reaction [59, 60]. Fig. 20 shows that Tg changes with step-wise post-curing for the phenolic-cured epoxy resins using the accelerators listed in Table 5. Even though 2 h at 200 °C is insufficient to cure the resins fully, the reactivity of functional groups is significantly decreased, which results in the saturation of Tg. More post-curing brings about resin degradation, but not higher Tg. It is interesting to note that the saturated

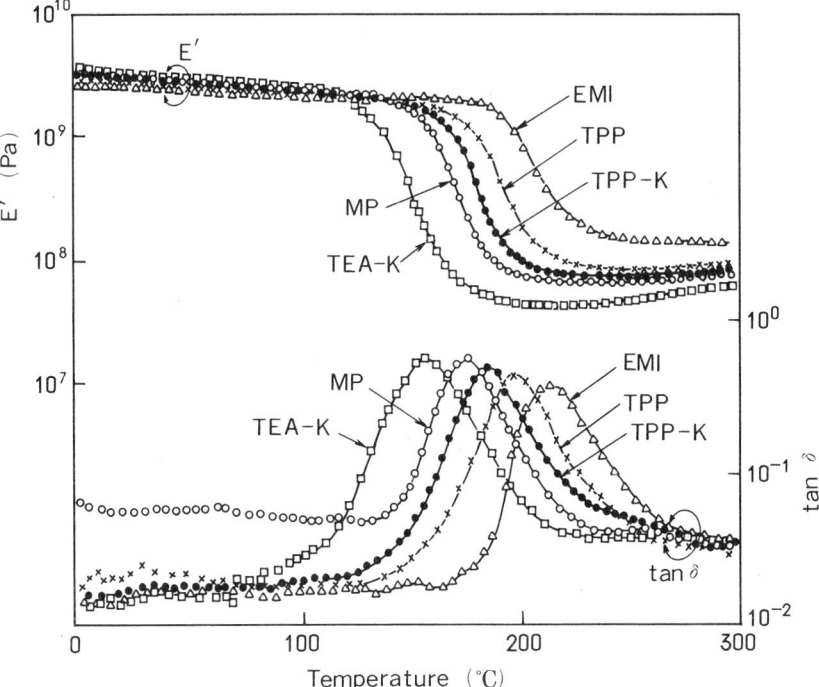

Fig. 21. Dynamic mechanical properties of cured epoxy resins

values of Tg differ depending on the accelerators used. Such a result points out that resins of different crosslinking densities may be obtained by using different kinds of accelerators. Viscoelastic behavior of such cured epoxy resins, as obtained by using different accelerators, is shown in Fig. 21. In the figure, the glass transition temperature region and elastic modulus E' of the rubbery region are seen to differ significantly depending on the accelerators used. This suggests that the accelerators influence not only the reaction rate, but also the physical properties of cured products, as discussed in the next section.

4.2 Influence of Crosslinking on Physical Properties

The most important characteristic of thermo-setting resins is that their chemical structures consist of molecular cross links, and crosslinking density, ϱ is one of the most important factors to determine physical properties of such resins. For example, it is known that the greater the crosslinking density ϱ becomes, the higher the glass transition temperature Tg becomes for resin polymers consisting of the same segments. The following relation usually holds between Tg and ϱ [58, 112],

$$Tg = K_1 \log \varrho + K_2 \qquad (8)$$

where K_1 is a constant denoting the degree of influence of cross-linking points on the molecular motion of chains. Its value differs greatly depending on the kind of thermo-setting resins, being found from 50 to 250. It is clear from Eq. (8) that the greater K_1 is, the greater the influence of the crosslinking points becomes on Tg, and consequently on molecular motion. Crosslinking density $\varrho(E')$ can be calculated from the rubbery elastic modulus E by using statistical theory for rubber elasticity [113].

$$\varrho(E') = \frac{E'}{3\psi RT} \qquad (9)$$

where ψ is a front factor, the value of which is usually 1; R is the gas constant; and T is absolute temperature.

Relations between the cross-linking density $\varrho(E')$ calculated from Eq. (9), and the glass transition temperature Tg are shown in Fig. 22. The fact that the higher the crosslinking density is, the higher Tg becomes satisfies Eq. (8). The optimum mixing ratio of epoxy resins and hardeners differs for the kind of hardening accelerators used [111]. This suggests that reactions of epoxy and phenol groups take place not necessarily on an equivalent basis, but concurrently between epoxy groups and other groups to form a three-dimensional network structure; then curing mechanisms also differ according to the kind of hardening accelerators [68, 114]. If this is true, then some residual functional groups will be left uncured at thermosetting [115]. Even when this thermoset resin is heated again at temperatures above Tg, for example, at 220 °C, its Tg will not change very much due to saturation. Rather, the cured resin is found to undergo thermal degradation when left for a long time. Therefore it seems that reactivity of functional groups is suppressed very much by steric hindrance and topological effects of molecular chains being formed around the functional groups, after being once cured and having the three-dimensional network form.

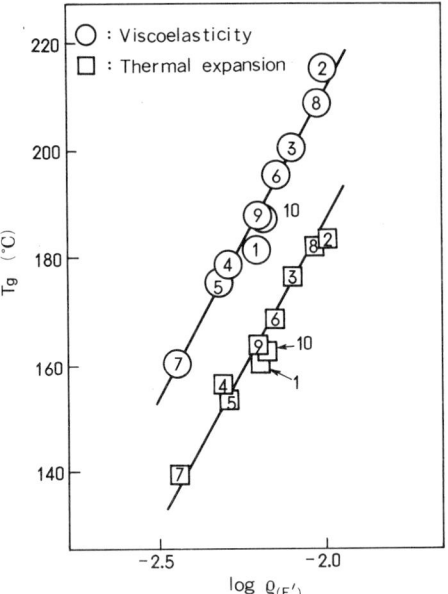

Fig. 22. Relationship between Tg and log ϱ(E') (numbers of the points refer to samples listed in Table 6)

Linear thermal expansion coefficients αL of thermoset resins in rubbery and glassy states are plotted against ϱ(E') in Fig. 23. A unique phenomenon is observed in the relation between αL and ϱ(E') which is reversed above and below Tg. That is, in the rubbery state of thermoset resins with higher ϱ(E'), αL tends to become smaller, while in the glassy state of thermoset resins with lower ϱ(E'), αL becomes larger.

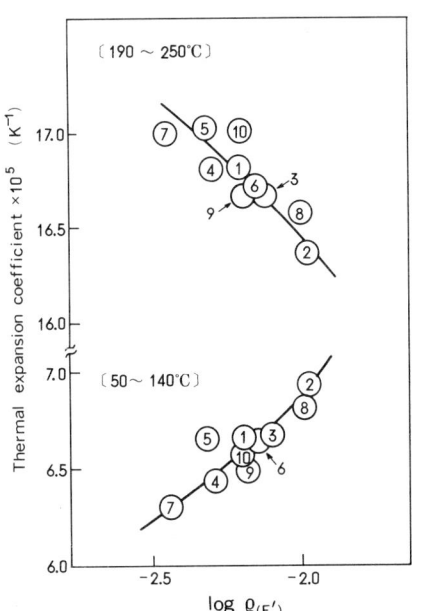

Fig. 23. Relationship between thermal expansion coefficient and log ϱ(E') (numbers of the points refer to samples as listed in Table 6)

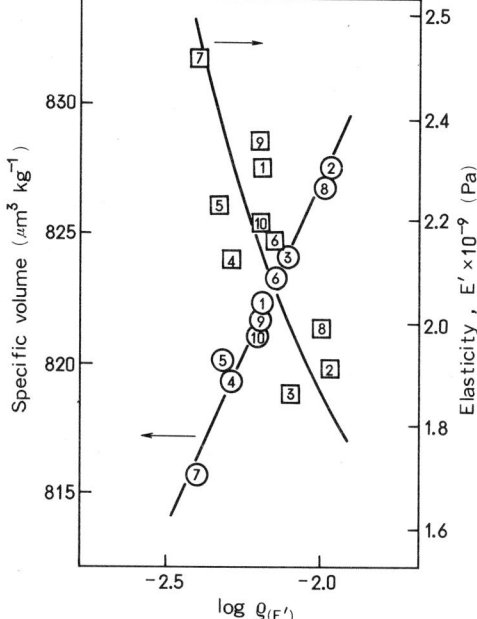

Fig. 24. Relationships between specific volume and elastic modulus, and log ϱ(E') (numbers of the points refer to samples as listed in Table 6)

Relations between specific volumes and elastic moduli, and crosslinking densities in the glassy state (lower than Tg by 60 °C) are plotted in Fig. 24. Specific volumes tend to increase with the increase of ϱ(E'), showing a decrease in density, while elastic modulus, to the contrary, tends to decrease. Such trends that densities for the glassy state and the elastic modulus tend to become smaller, when crosslinking densities of thermoset resins (i.e. Tg) become higher, are not unique to only these systems, but are observed in many other systems [58-62]. However, this may seem hard to understand and even paradoxical when conventional thinking indicates that higher crosslinking density brings about higher stiffness and lower specific volume. What must be noted first, however, is that such phenomena are behavior in glassy states where Brownian motions of molecules are frozen. It may be assumed, therefore, that crosslinking points have some specific effect on the initiation of the glassy state, as well as some effect to loosen packed states of molecular chains. Thus, with an increase in crosslinking densities, the distance between molecular chains becomes and wider, which may be considered to result in decreases in macro densities and in the elastic moduli.

Absorption, diffusion and permeation of water through the epoxy molding compound is directly related to the reliability of moisture resistance of molded devices, as was discussed in Sect. 2.2 [116]. Then here only the moisture-proof properties of epoxy resins themselves in the cured state are mentioned. The relationship between such water-resistance properties and crosslinking density of cured epoxy resins is shown in Fig. 25. Higher crosslinking density promote water absorption, diffusion and permeation through the cured epoxy resins, which is undesirable as far as molding compound reliability is concerned. However, considering that water molecules can pass through loosely packed intermolecular space or free volume, this

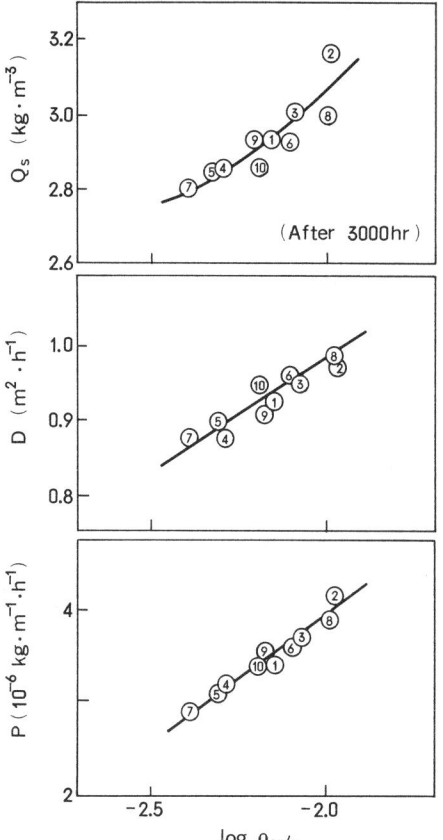

Fig. 25. Effects of $\varrho(E')$ on the saturated water content (Qs), water vapor permeability (P) and diffusion coefficient (D) of cured epoxy resins at 60 °C, 95% RH (numbers of the points refer to samples as listed in Table 6)

tendency seems consistent with the result in Fig. 25, which shows that higher crosslinking of epoxy resins brings about low macro densities. That is, since molecular chains are loosely packed in the highly crosslinked resins, water molecules seem to be able to penetrate and move more easily through the resin matrix. In the following section, such abnormal phenomena of Fig. 23 to 25 are related to each other.

4.3 Interpretation of Curing and Thermal Shrinkages

In Fig. 26, changes of specific volumes of thermoset resins undergoing curing and thermal shrinkage are schematically shown. Resins I and II are characterized by high and lower Tg values which result from accelerators giving high and lower $\varrho(E')$ values, respectively.

Resin I shows a large shrinkage on curing, resulting in its high $\varrho(E')$ and small specific volume in the rubbery state of the cured resin, while resin II shows a comparatively small shrinkage on curing, resulting in its small $\varrho(E')$ and large specific volume. As for the thermal expansion coefficients in the rubbery states, resin I with higher $\varrho(E')$ has a smaller value. At temperatures higher than Tg, polymers are held in a

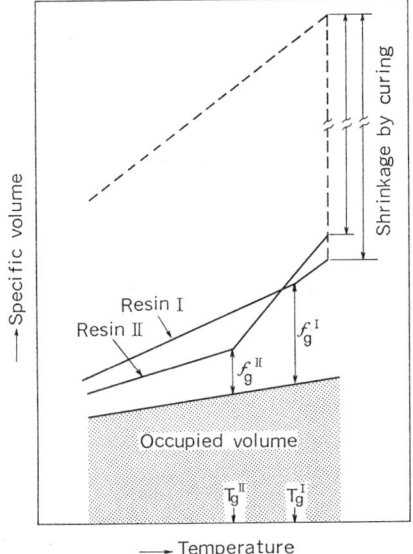

Fig. 26. Schematic representation of specific volume change of epoxy resins during curing

thermal equilibrium condition, allowing active micro-Brownian motions for the molecular chains. The crosslinking points in the rubbery state of a three-dimensional network structure are considered to restrain micro-Brownian motions of molecular chains in the resins. Therefore, thermal expansion can be interpreted as being restrained as much as the crosslinking density is increased, because of the mutual pulling and restraining of molecular chains. However, the tendency for thermal expansion coefficients to become smaller for thermoset resins with greater $\varrho(E')$ in the glassy state cannot be explained by the above thermal equilibrium theorem. Therefore, it should be considered that frozen-in micro-Brownian motions in the two resin systems differ because of the difference in their crosslinking densities.

As a next step, the cooling process of thermoset resins after curing is examined. In resin I, micro-Brownian motion will be frozen in at a comparatively large free volume because its motion is already restrained to some degree by crosslinking points due to the high $\varrho(E')$. On the other hand, in resin II, its micro-Brownian motion starts to be frozen in when its free volume becomes smaller than that in resin I because of the smaller restraining effect of crosslinking points due to the smaller $\varrho(E')$. Thus, in the glassy state, the normal thermal shrinkage on cooling is inhibited by the crosslinking point, resulting in higher specific volume (or lower density), and smaller modulus of elasticity in resin I than in resin II. Since the thermoset resins used here were prepared from the same epoxy resin components, except for the accelerators, their occupied volumes are approximated by being put equal. Thus, the difference in specific volumes can be attributed to that in the free volume. Thermal expansion of materials consists of thermal expansion of its occupied volume and free volume. Thus, the reason why resin I with high $\varrho(E')$ and Tg has a larger thermal expansion coefficient in the glassy state may be attributed to its larger free volume.

According to the theory of the iso-free-volume [116], a value of fractional free volume fg at Tg is said to be 2.5% regardless of the kinds of polymer materials. However, epoxide resins with three dimensional structures are considered to have different fractional free volumes when becoming glassy, because restraining effects on micro-Brownian motions will differ in the numbers of crosslinking points or how they are bridged.

5 Development of Highly Reliable Encapsulation Materials

Integration density of LSIs has been increasing very rapidly, and 4 Mbit DRAMs are expected to be marketed in 1989 [54, 117]. With such a high-pitched increase in integration, finer patterning and greater multilevel interconnections have been adopted for the wiring of devices, gradually yielding larger-sized chips [122]. Even though the chip sizes increase, compacter packages are required to realize the higher integration, which results in thin-skinned plastic encapsulation. Moreover recent developments in direct surface mounting technology necessitates that the encapsulation materials withstand soldering temperatures [52-54]. Then, requirements for encapsulation materials are becoming stricter either as demands for improved characteristics or satisfying newly added performance needs. Development of high quality, high performance molding compounds is desired, because molding compounds have been and will be the mainstay of encapsulation materials for LSIs, in view of cost and productivity.

Important characteristics required for encapsulation materials, as discussed in Sect. 2, are summarized in the following.

1) Low Thermal Stress [9, 17-20, 54, 83, 84]

Thermal stress in encapsulation materials should be intrinsically small in order to prevent occurrence of cracks in packages or passivations layers, displacement of aluminum wirings, short circuits between layers, or changes in device performance, caused by the pile-up of thermal stress in the package. This becomes particularly acute with increases in packaging density.

2) Moisture Resistance [9, 29-47]

A high level of moisture resistance must be given to prevent corrosion of aluminum wiring, which tends to be observed more often for thin-skinned encapsulation and finer patterns for wiring. Thermal shock at surface mounting on printed wiring boards sometimes causes cracks in packages, resulting in a decrease of moisture resistance.

3 High Heat Resistance [47, 48, 100-103]

When encapsulated components are exposed to temperatures above 200 °C for long times, contacts between gold and aluminum inside the components may be corroded, impairing conductivity. As the thermal decomposition properties of the resins used have much to do with such corrosion, heat resistance of the encapsulation materials must be improved.

4) Moldability [3, 64, 76]

Desirable properties, such as easy filling and easy release, void-free, and warp-free molding and no staining of the molds must be assured for the molding resins to be used.

5) Elimination of α-Rays [49-51]

Efforts to eliminate α-ray sources in encapsulation materials have become necessary because, with the decrease in cell areas, the soft error of devices from the α-rays has become serious.

5.1 Lowering Thermal Stress in Molding Compounds

Thermal stress in resin encapsulated LSI packages is produced by the differences in thermal expansion coefficients between encapsulation materials and the insert materials such as silicon chips, lead frames, etc. The thermal stress σ generated can be estimated by the following equation, which is approximately derived from Eq. (1).

$$\sigma = k \int_{T_1}^{T_2} (\alpha_p(T) - \alpha_c(T)) E_p(T) \, dT \tag{10}$$

where k: constant; $\alpha_p(T)$: linear thermal expansion coefficients of encapsulation materials; $\alpha_c(T)$: linear thermal expansion coefficients of chips; and $E_p(T)$: modulus of elasticity of resins. Generally, $\alpha_p(T)$ differs depending on the temperature, while $\alpha_s(T)$ has hardly any dependence on the temperature, keeping a constant value.

Linear thermal expansion coefficients and elastic moduli of encapsulation materials change greatly at temperatures near Tg, however, they do not change very much above or below Tg. Therefore, Eq. (10) can be reduced to the following;

$$\sigma = k\{(\alpha_1 - \alpha_s) E_1(T_g - T_1) + (\alpha_2 - \alpha_s) E_2(T_2 - T_g)\} \tag{11}$$

where α_1: linear thermal expansion coefficient of resins applicable at temperatures lower than Tg; α_2: those at temperatures higher than Tg; E_1: elastic modulus at temperatures lower than Tg, E_2: that at temperatures higher than Tg. Eq. (11) suggests that the following three methods are useful for lowering thermal stress in the encapsulated devices.

1) Lower the linear thermal expansion coefficient of molding compounds to those of inserts.
2) Lower the elastic modulus of molding compounds.
3) Lower the Tg of molding compounds.

Considering that high mechanical strength of molding compounds is significantly required at the elevated temperatures of reflow soldering, the third method to decrease the thermal stress, that is, "lowering the Tg", is not suitable because a low Tg for thermosetting resins generally results in low mechanical strength at elevated temperatures.

Most of the low thermal stress molding compounds now on the market are based on epoxy resins having rubber particles dispersed as flexibilizers to lower the elastic modulus in the cured state [28, 54, 85, 86]. Although this method has been successful so far, it will not to lead any further lowering of the thermal stress, a requirement which is becoming greater and more important in VLSI applications. Further effort to prepare molding compounds with their inherent thermal stress remarkably decreased is being made by matching thermal expansion coefficients between molding compounds and inserts. Increasing the amount of fillers to be added is effective for lowering thermal expansion coefficients as shown in Fig. 17, but is consequently degrades resin moldability because viscosity is increased. To deal with such problems, development of fillers that cause no viscosity rise in resins, or addition of flexibilizers must be considered. Various relationships have been proposed and discussed regarding the influence of fillers on viscosity in encapsulation resins, especially where a large quantity of fillers is used. Mooney's viscosity Eq. (12) [119], however, explains it well,

$$\ln \eta r = \frac{Ke\varphi}{1 - \varphi/\varphi m} \tag{12}$$

where ηr: relative viscosity; Ke: Einstein coefficient (depending on shapes and aggregation of grains); φ: volume fraction of fillers; φm: the maximum volume fraction allowed for fillers (depending on particle size distribution, shapes or aggregation of grains). Ke in Mooney's equation is dependent largely on the shape of fillers, and takes smaller values when spherical fillers are used rather than crushed ones having angular shapes. φm is dependent largely on the particle size distribution of fillers, and takes smaller values for wider distributions.

In compliance with the above formula, spherical fillers have been used by expanding and controlling the grain size distribution in order to prevent increased viscosity of the resin compounds when the quantity of added filler is increased. Recently, mass-produced spherical powders of fused silica are being used widely for various purposes such as lowering viscosity and increasing fluidity of the resin compounds before curing, and improving thermal expansion coefficients or thermal conductivity to disperse thermal stress in molding materials in the cured state [54]. Shapes of conventional crushed filler and the newly developed spherical one are compared in Fig. 27.

Fig. 28 relates viscosity of epoxy molding compounds and filler content. The particle size distribution is varied over a wide range, which is represented by the n value of the Rosin-Rammler equation [93]. From Fig. 28 the decrease in the n value, which means a broader distribution of particle sizes, results in a decreased viscosity of the molding compounds. The same tendencies are observed for both spherical or crushed fillers. With regard to the effects of filler shape on the viscosity, spherical filler is much more effective for decreasing the viscosity than is crushed. In conclusion, by adopting spherical filler with an optimized grain distribution, the amount of added filler can be increased without increasing resin viscosity. By adding a specific silicone resin as a flexibilizer, the elastic modulus and the linear expansion coefficient of the cured resin can also be decreased at the same time. Therefore, the thermal stress of the cured resin can be remarkably decreased by applying these two methods, and balancing the various requirements and characteristics for the encapsulation materials.

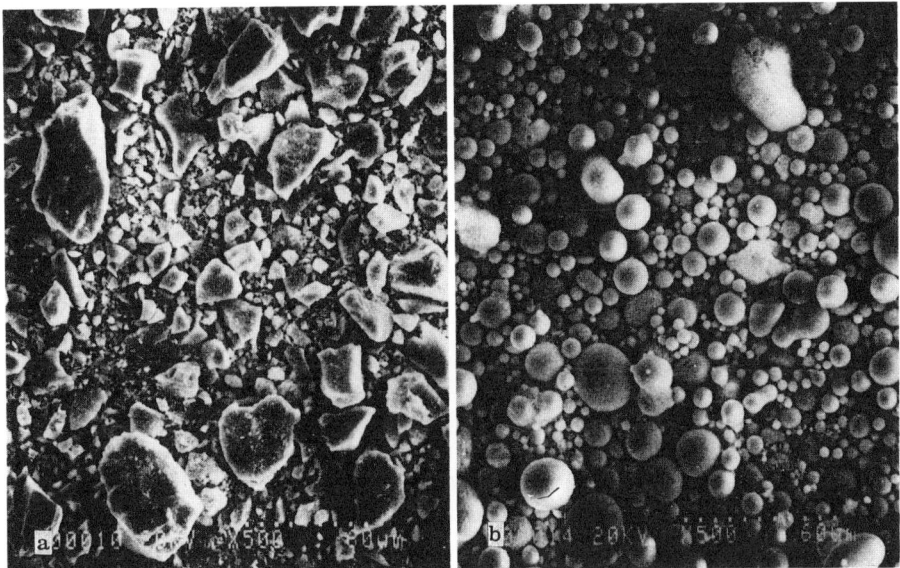

Fig. 27a and b. SEM photograph of fused silica fillers

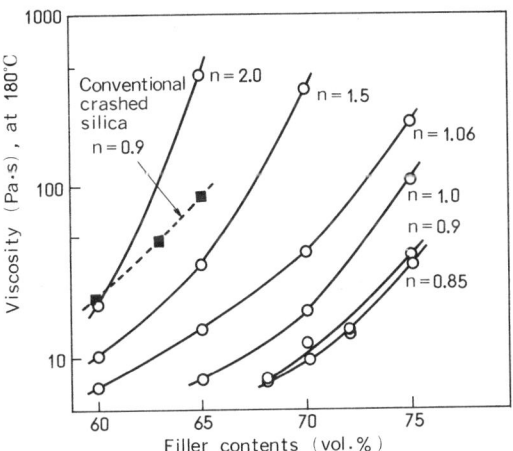

Fig. 28. Effects of the grain size distribution of the spherical silica fillers on the viscosity of the epoxy molding compounds. (n is an index in Rosin Rammler equation and indicates grain size distribution. Smaller n broader grain size distribution and larger n narrover grain size distribution.

5.2 Moisture-Proof Reliability

Most of the problems concerning reliability of moisture resistance are attributable to corrosion of aluminum wirings [9, 29, 30, 33, 34]. Moisture or ionic impurities such as Cl^-, Br^-, $RCOO^-$, Na^+, etc. included in the bulk materials or penetrating through interfaces of the encapsulation materials and lead frames are considered to be the major cause of corrosion of aluminum wirings [38, 39]. Needs for a larger scaled chip, thin-skinned encapsulation, surface mounting, and higher temperature packaging processes, when eventually combined together, will further allow penetration of moisture or impurities from the outside, increasing corrosion.

First of all, it is necessary to obtain higher purity components. For this purpose, not only must be the epoxy resins chloride ion-free, but they must also be phenol-free. Ionic impurities in phenol novolac resin hardners, and trace radical bromine and chlorine in brominated epoxy resins, antimony trioxide or coloring agents should be kept to a minimum. Next, ratios of coupling agents and release agents must be evaluated to determine the best combination yielding the optimum adhesion between encapsulation materials and devices or lead frames.

The main type of package for resin encapsulated devices is going to change from insertion mounted dual in-line packages (DIP) to surface mounted packages [117]. It has already been recognized that moisture resistance reliabilities of molded devices sometimes degrade after reflow soldering, because packages warp or crack and delamination takes place inside them at the surface mounting on printed wiring boards [52, 53]. In order to prevent such problems, efforts to develop high temperature, high strength molding compounds have been made, according to the suggestions pointed out in Sect. 2. The key to develop such molding compounds should be found in reading higher molecular weight and multifunctional epoxy resins or hardners as well as applications of new hardners, because highly crosslinked resins can provide excellent stability at elevated temperatures of reflow soldering [54].

Recent low thermal stress encapsulation materials have been obtained by integrating many related advancements in materials science and engineering to achieve high purity, high adhesive and high heat resistance materials, with technology to lower thermal stress. Low thermal stress materials show excellent moisture resistance reliability even after solding.

5.3 High Heat Resistance

The reliability of resin molded semiconductor devices has been remarkably improved, and their increased applications to communication services, utilities and industrial equipment are expected. As one of the reliability tests to qualify for such applications, accelerated lifetime tests of resin molded semiconductor devices have been conducted at temperatures above 200 °C, and their reliability when exposed to high temperatures for long periods has been evaluated. Normally, Tg of conventional mold resins lies in a temperature range of 150–180 °C, and, thus, there are few data on reliability tests above 200 °C [47, 48]. While more data for the accelerated tests at above 200 °C must be obtained before validating results, it has been shown that an electric conduction decreases between Al electrodes and Au wirings [100–103, 118]. The entire package is exposed to high temperatures of 200 ~ 300 °C on soldering, though for a short time, during the surface mounting process. It is, thus, troublesome that bonding between Au and Al may suffer damage with such exposure, and device reliability may be degraded.

As flame retardant' brominated epoxy resins promote corrosion of Au/Al bonding [47, 48, 101–103], it is advisable to decrease their amount and, instead, to increase the amount of antimony trioxide used. A mechanism for high temperature degradation of Au/Al bonding is considered to be that intermetallic compounds formed at the bonding interface are corroded by halogen compounds from thermal decomposition of mold resins.

5.4 Lowering α-Ray Sources

Soft error troubles have become very important since the introduction of 64 k or 256 kbit DRAMs in which memory cell areas had been greatly reduced [49]. Soft errors are caused by α-rays emitted from radioactive impurities such as uranium or thorium included in encapsulation materials [50,51]. Most of the α-rays from the encapsulation materials, however, are attributed to fused silica used as fillers as shown in Table 6 [120,121]. Therefore, using high purity silica with less inclusion of U and Th is one possible solution. Another solution is to coat chip surfaces with polyimides or silicone resins because these polymer films have shielding effects against α-rays [123]. Both solutions has been adopted on up to 256 Kbit DRAMs [117]. However, both have some merits and disadvantages. That is, when high purity silica is used, the cost of encapsulation materials becomes very high. Chip coating requires complicated processes, and a film thickness of 50 μm to guarantee sufficient α-ray shielding, which introduces degradation and loss of reliability due to the differences in adhesion or mismatches in thermal expansion coefficients between coated films and chips, molding materials, or Au wirings. Chip coating, however, is advantageous in that it mechanically protects device surfaces on which finer patterned and more multi-leveled integrated circuits are being mounted. A protective technique combining the above two methods is to be employed for 1 Mbit or larger DRAMs in the future.

Table 6. α-Ray emission level of raw materials and encapsulation materials

Samples		U content (ppb)	α-ray flux ($\alpha/m^2 \cdot h$)
Raw materials	Epoxy resin	<1	50
	Brominated epoxy resin	<1	100
	Hardener	<1	50
	Filler (Conventional grade)	80	6300
	Filler (Special grade)	0.3	370
	Flame retardant (Antimony trioxide)	20	400
Encapsulation materials	Conventional grade	—	340
	low α-ray flux grade	—	4

6 Rheological Characteristics and Mold Filling Dynamics of Epoxy Molding Compounds

As stated in Sect. 5.1, there are many trade-off points to be considered in the requirements for the properties of encapsulation materials. One particularly important example is melt viscosity of molding compounds which increases when the filler content is increased in order to meet the requirement for lowering thermal stress in packages. Recent epoxy molding compounds actually have poorer moldability as compared with convensional molding compounds. This poorer moldability may cause defects in the packages, such as wire sweep and void formation etc., if adequate measures are not taken. Therefore, it is very important to develop a method for analyzing rheological characteristics and mold filling dynamics of epoxy molding compounds, and to develop a new mold design system by which the goal is decreasing defects in packages.

6.1 Flow Characteristics of Epoxy-Filler Systems

While viscosity of thermoplastics is dependent primarily on the parameters of temperature and shear rate, thermosetting resins are accompanied by an increase in viscosity due to chemical reaction at elevated temperatures. This behavior causes a difficulty in formulating mathematical models to analyze the molding process of thermosetting resins theoretically. Consequently, many empirical flow tests, such as the spiral flow test of EMMI I-6, etc., have been developed and widely used for inspection of moldability of thermosetting resins [124]. However, the results obtained by these methods do not give useful information for prediction of mold filling behavior in the actual molding process. Most actual molds have larger and varied dimensions of the flow channel than those of the test molds. In other words, there is a great difference in the rheological behavior between the practical and testing process. Therefore, it is customary for resin makers to have their own molds for IC encapsulation in order to evaluate the practical moldability of molding compounds. However, the test methods are useful only for relative comparisons of flow characteristics of the molding compounds under limited conditions. This fact makes the production management of IC encapsulation complicated and time consuming.

Empirical models to describe viscosity variations with lapsed time and elevated temperature have been proposed for the case of epoxy resins by some workers [125, 126]. These isothermal models, however, seem to be inadequate, since the viscosity does not reach infinity within the gel time of the resin, according to these treatments.

The other models have been proposed by recent workers to correlate viscosity variations to chemical curing states [127-130]. These approaches are valuable to characterize the rheological properties of thermosetting resins in a theoretical manner. However, in order to obtain detailed information on the reaction kinetics and molecular weight variation as seen in their treatments, it is necessary to have detailed information on each component in the molding compounds and on their thermal histories during their material processing such as kneading. Such information usually is unavailable in practice for the commercially available molding compounds.

A simple and useful treatment has been developed on the viscosity variation for isothermal and nonisothermal reactions of commercially available thermosetting molding compounds, which provides a way to simulate the mold filling behavior of thermosetting resins in the flow channels of actual molds used for IC encapsulation [131]. The viscosity is treated as a function of temperature and time, independent of shear rate, based upon the fact that the shear rate dependence of epoxy resin compound may be assumed to be negligibly small as compared with the temperature dependence.

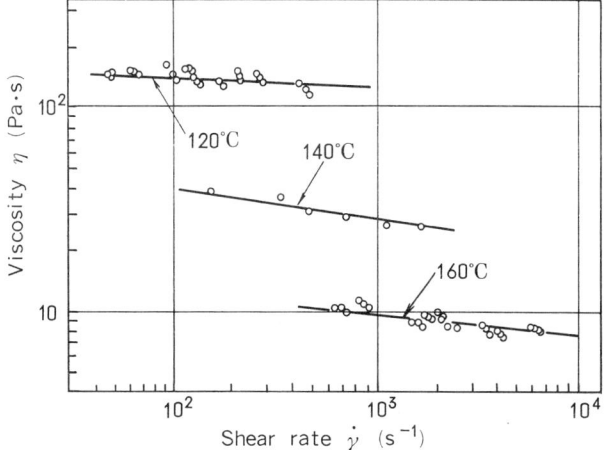

Fig. 29. Shear viscosity vs. shear rate for the trial resin system (silica filled epoxy resin compound having neither curing agent nor catalyst) at 120, 140 and 160 °C, measured by a plunger extrusion type of rheometer

To confirm this assumption, the shear viscosity of a silica filled epoxy resin systems, in which hardener and accelerator are both absent so that no chemical reaction is involved, was examined with temperature as a parameter. Results are shown in Fig. 29. The data show that the viscosity at each temperature decreases slightly with increasing shear rate over the range of actual molding conditions. On the other hand, it is obvious that the temperature difference produces a more drastic change in viscosity, however. From this experimental result, it may be assumed at least during the first stage of curing (in other words, during the mold filling process) that the viscosity is a function of temperature and time, and independent of shear rate.

On the basis of this assumption, an empirical model was proposed to describe isothermal viscosity change of the epoxy resin compounds during their mold filling process.

$$\eta = \eta_0(T)\,[\{1 + t/t_0(T)\}/\{1 - t/t_0(T)\}]^{C(T)} \tag{13}$$

where η: viscosity
η_0: initial viscosity at zero time
t: lapsing time
t_0: gel time

Eq. (13) satisfies the following boundary conditions:

$$\eta = \eta_0(T) \quad \text{at} \quad t = 0, \quad \eta = \infty \quad \text{at} \quad t = t_0(T) \tag{14}$$

η_0, t_0 and C can be expressed as:
$\eta_0 = a \exp(b/T)$;
$t_0 = d \exp(e/T)$;
$C = f/T - g$

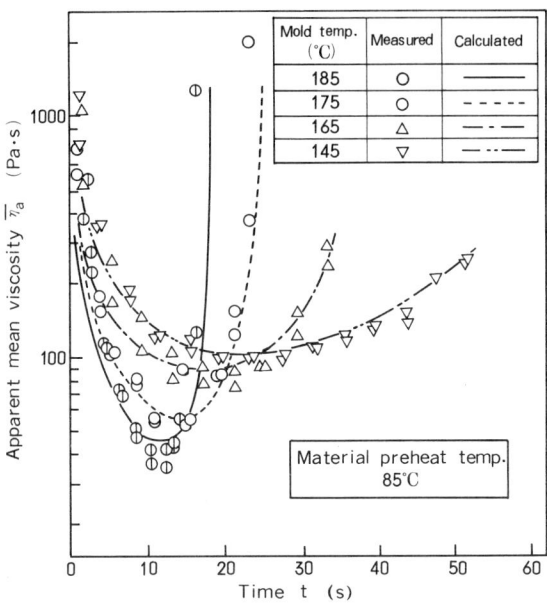

Fig. 30. Comparison of measured and calculated profiles of the apparent mean viscosity vs. curing time of epoxy molding compound for IC encapsulation

a, b, d, e, f, and g are constants inherently related to the nature of the epoxy resin compound itself.

Successive calculations of viscosity as a function of temperature and lapsed time provide the viscosity change for the nonisothermal process. A simultaneous equation, involving the conservation equations for the total mass, momentum and energy balance of the resin in the flow channel, and the above model, Eq. (13) have been successfully introduced to compute the apparent mean viscosity of the commercial epoxy resins in a circular cross-sectional flow channel with the aid of a FEM (Finite element method). The constants, a to g, were determined by means of curve fitting of the calculated apparent mean viscosity to experimental data. Fig. 30 shows that the calculated profiles of the variation in the apparent mean viscosity fit those of the experimental data. These constants thus determined have been used for the simulation of viscosity change and pressure drops of the epoxy molding compounds during the mold filling in the transfer molding paths of actual molds.

A Resin rheological parameter analyzer (RPA) has been developed, consisting of a transfer molding machine, test molds having a circular cross-sectional flow channel, and data processing and display units. This system provides basic rheological properties of the resin to be inspected such as the apparent mean viscosity and melt front distance vs. time. The apparent mean viscosity of the epoxy molding compound filled with spherical silica grains has been compared with that filled with the conventional crushed silica grains, as shown in Fig. 31. It is obvious that the moldability of the epoxy molding compound with spherical silica grains is improved greatly as compared to that with the crushed silica grains. RPA can be utilized to inspect moldability of resin compounds in the manufacturing process, providing more reliable data than

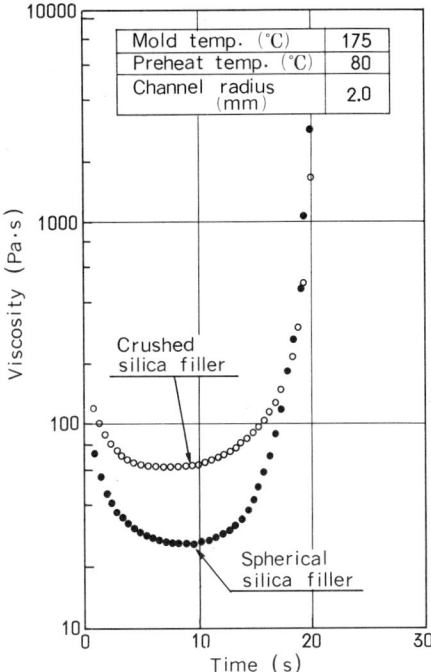

Fig. 31. Comparison of the profile of the apparent mean viscosity vs. time for the silica filled epoxy molding compound, measured by RPA. (○ denotes conventional crushed fused silica and ● denotes spherical fused silica)

the conventional spiral flow test. The combination of data obtained from RPA and the mold filling simulation model described above has been utilized to inspect the moldability of various kinds of actual molds successfully.

6.2 Uniform Mold Filling Method

A computer program for the design of flow channels has also been developed to get uniform mold filling. This program is known as runner and gate variation (RGV). The resin melt can fills all the cavities at almost the same velosity. The program compu-

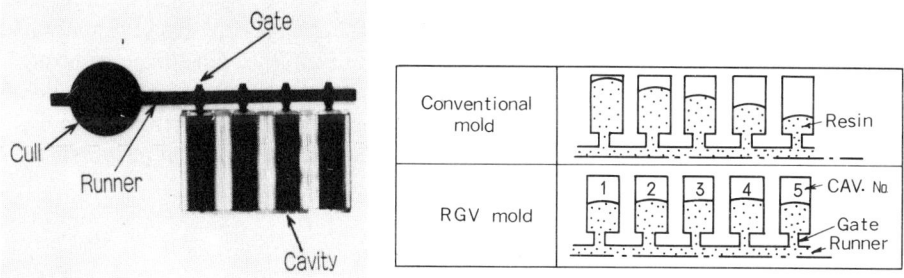

Fig. 32. Comparison of the mold filling characteristics of the RGV mold with those of the conventional mold

tes the dimensions of the runner and gates using the expanded simulation program and the constants of Eq. (13), in the manner of balancing the pressure drops for all the transfer molding paths. The mold dies designed by RGV (RGV mold) consists of a runner having the convergent angle and gates for which the convergent angle are successively increased in the direction of the down flow [132].

The mold filling characteristics of the RGV mold have been compared with those of the convention mold, and are shown in Fig. 32. The filling behavior of the RGV mold changes slightly, depending upon the resin used and the molding conditions, but it has more uniform filling behavior than conventional molds. The RGV mold has an advantage of decreasing mold defects such as void formation and insert deformation, as compared with conventional molds.

7 New Approachs to Encapsulation Materials

The major requirements of encapsulation materials are higher productivity and lower cost for small chips on one hand, including bipolar and discrete devices, and low thermal stress, higher heat resistance and solder heat resistance for large chips in SMDs on the other hand, including memory, microcomputers and logic devices. More than 10 years ago PPS (poly-p-phenylene sulfide) was first investigated for use in semiconductor encapsulation, and recently an improved material has been obtained and reinvestigated. Polyimides have also been proposed as encapsulation materials, because of their excellent thermal stability. These materials are discussed below [54].

7.1 Poly-p-Phenylene Sulfide

Many thermoplastic polymers have been investigated for semiconductor encapsulation because such polymers can be molded by injection machines without post-curing, contributing to excellent productivity. Among them, PPS has been examined for its higher heat resistance and low melt viscosity. Applications are restricted to discrete devices such as diodes and transistors, because they require productivity (i.e. fast cure and low cost) first, and their chip sizes are so small that it is easy to maintain reliability. Since yields of the encapsulation materials in the transfer molding process are about 20 to 30% for those for small packages, more than 70 to 80% of the materials are thrown away when used thermosetting resins. But injection molding can save materials because their recycling is possible. Therefore, there is still an advantage of cost in applying PPS to encapsulation materials, even though its price is twice that of epoxy molding compound. The drawback of using PPS for encapsulation materials are high viscosity causing wire sweep at molding and package deformation of at reflow soldering.

7.2 Polyimides

Another approach to semiconductor encapsulation is using polyimides. Polyimides are promising because of their higher glass transition temperature, low coefficient of thermal expansion, and higher strength at elevated temperatures. Although polyimides have good characteristics, there are still a few problems to be solved. Two are curing temperature and time. Current materials require curing time of 5 min at 200 °C, and 5 h at 200 °C for post-curing, but these curing conditions are undesirable from the viewpoint of productivity. A third problem is large moisture absorption by polyimides. This requires that the polymers themselves must be improved and/or some modifications made.

8 Summary

The major advantage of microelectronics device encapsulation by using polymers is economy in production where many parts can be packaged simultaneously with a relatively low cost material. The history of plastic package technology has involved meeting the challenge of achieving better product reliabilities. The requirements for encapsulation materials have been changing almost yearly, becoming more and more severe as trends show in Fig. 33. The future of these materials will continue to require development of new processes to provide better products for tomorrow's user.

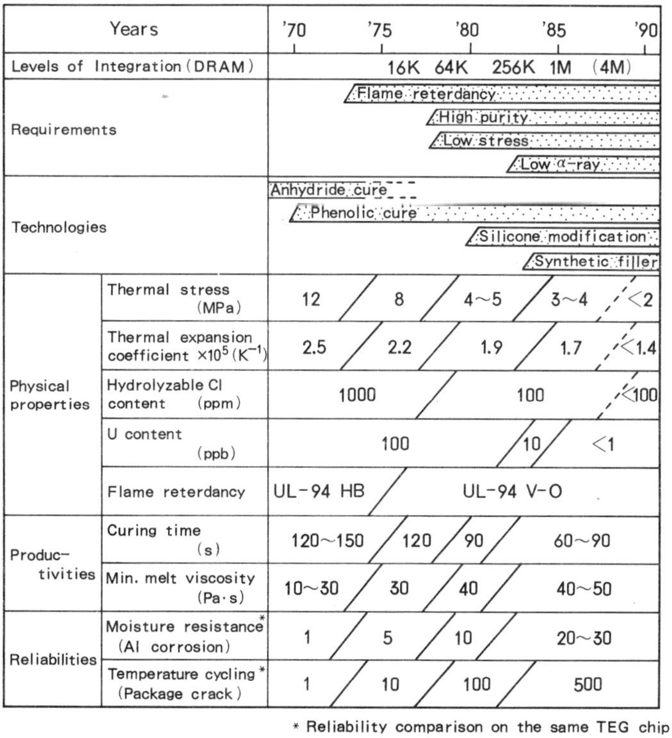

	Years	'70	'75	'80	'85	'90
Levels of Integration (DRAM)			16K	64K	256K 1M	(4M)
Requirements			Flame retardancy	High purity	Low stress	Low α-ray
Technologies			Anhydride cure			
			Phenolic cure			
					Silicone modification	Synthetic filler
Physical properties	Thermal stress (MPa)	12	8	4~5	3~4	<2
	Thermal expansion coefficient ×10⁵ (K⁻¹)	2.5	2.2	1.9	1.7	<1.4
	Hydrolyzable Cl content (ppm)		1000		100	<100
	U content (ppb)			100	10	<1
	Flame retardancy	UL-94 HB		UL-94 V-O		
Productivities	Curing time (s)	120~150	120	90	60~90	
	Min. melt viscosity (Pa·s)	10~30	30	40	40~50	
Reliabilities	Moisture resistance* (Al corrosion)	1	5	10	20~30	
	Temperature cycling* (Package crack)	1	10	100	500	

* Reliability comparison on the same TEG chip

Fig. 33. Trends in epoxy encapsulation materials for transfer molding and related reliabilities

9 References

1. Peck DS (1970) 8th Ann. Proc. International Reliability Physics Symposium, p81
2. Brauer JB, Kapfer VC, Tamburrino Al (1970) ibid. p61
3. Bevinton JR, Cook JP, Little DR (1970) ibid. p73
4. Yoshida T, Takahashi T (1982) 20th Ann. Proc. International Reliability Physics Symposium, p268
5. Gunn JE, Mazumdar PM (1979) 17th Ann. Proc. International Reliability Physics Symposium, p48
6. Peeples JW (1978) 16th Ann. Proc. International Reliability Physics Symposium, p154
7. McGarvey WJ (1979) 17th Ann. Proc. International Reliability Physics Symposium, p136
8. Timoshenko S (1955) Strength Materials. 3rd ed., Van Nostrand
9. Inayoshi H., Nishi K., Okikawa S, Wakashima Y. (1979) 17th Ann. Proc. International Reliability Physics Symposium, p113
10. Kuwata K., Ikoh K, Tabata H (1985) IEEE Trans. on Components, Hybrids, and Manufacturing Technology CHMT-8(4), p486
11. Spencer JL, Schrone WH, Bendnarz GA, Bryan JA, Metzgzr DR (1981) 19th Ann. Proc. International Reliability Physics Symposium, p74
12. Mason WP, Thurston RN (1957) J. Acourst. Soc. Am. 29: 1096
13. Miura H, Nisimura A, Kawai S, Nisi K (1987) Preprint for the 64th Ann. Meeting of JSME, p1821
14. Konda Y (1982) IEEE, Trans. Electron Devices, ED-29: 64
15. Tufte ON, Chapman PW, Long D (1962) J. Appl. Phys. 33: 3322
16. Shinohara M (1982) Plastics in Telecommunications III: 61
17. Okikawa S, Sakimoto M, Tanaka M, Sato T, Toya T, Hara Y (1983) International Symposium on Test and Failure Analysis, p1
18. Okikawa S, Suzuki H, Mikino H (1984) ibid. p180
19. Thomas RE (1985) 35th Ann. Proc. Electric Component Conference, p37
20. Shirley CG, Brish RC (1987) 25th Ann. Proc. International Reliability Physics Symposium, p238
21. Matsumoto H, Yamada M, Fukushima J, Kondoh, T, Kotani N, Tosa M (1985) 23rd Ann. Proc. International Reliability Physics Symposium, p180
22. Nishimura A, Tatemichi A, Miura H, Sakamoto T (1987) 37th Ann. Proc. ECC, p477
23. Yamada Y, Matsumoto H, Fukushima J, Kondoh T (1985) 23th Ann. Proc. International Reliability Physics Symposium, p192
24. Sasaki S, Serizawa K, Kaneda A, Nishi K, Hashizume S (1986) 5th International Conference on Reliability and Maintainability, p482
25. Yasukawa A, Sakamoto T, Shida S (1983) Proc. IEDM 83, p259
26. Mukai J, Kinjo N (1981) Practical Polymers for Engineers. Kodansha Scientific, Tokyo (in Japanese)
27. Paris PC, Erdogan F (1963) Trans. ASME, Series D, 85: 528
28. Shoraka F (1986) 6th Ann. International Electronics Packaging Conf., Proc. of the Technical Conf., p294
29. Comizzoli RB, White LK, Kern W, Schnable GL, Perters DA, Tracy CE, Vibronek RD (1980) 18th Ann. Proc. International Reliability Physics Symposium, p282
30. Berg HM, Paulson WM (1977) J. Electrochem. Soc. Ext. Abstr. 77-1: 33
31. Sato N (1971) Electrochemca Acta, Pergamon Press, 16: 1683
32. Caldwell BP, Albano V (1939) J. Trans. Electrochem. Soc. 76: 271
33. Deltombe E, Vanleugenhaghe C, Pourbaix M (1966) Pergamon Press, Oxford, p168
34. Tsubosaki K, Wakashima Y, Nagashima N (1983) 21th Ann. Proc. International Reliability Physics Symposium, p83
35. Fisher F (1970) 8th Ann. Proc. International Reliability Physics Symposium, p94
36. Sbar NL, Kozakiewticz RP (1978) 16th Ann. Proc. International Reliability Physics Symposium, p161
37. DerMardderosian A, Gionet V (1978) ibid. p179
38. Paulson WM, Lorigan RP (1976) 14th Ann. Proc. International Reliability Physics Symposium, p42
39. Kolser SC (1974) 12th Ann. Proc. International Reliability Physics Symposium, p155
40. Van de Ven EPGT, Koelmans H (1976) J. Electrochem. Soc. 123: 143

41. Paulson WM, Kirk RW (1974) 12th Ann. Proc. International Reliability Physics Symposium, p172
42. Koelmans H (1974) ibid. p 168
43. Altenpohl D (1957) Z. Metallk. *48*: 306
44. Altenpohl D (1953) Alluminum *29*: 361
45. Wakashima Y, Inayoshi H, Nishi K, Nishida S (1976) 14th Ann. Proc. International Reliability Physics Symposium, p223
46. Shockley W (1964) Surface Science *2*: 277
47. Gale RJ (1984) 22th Ann. Proc. International Reliability Physics Symposium, p37
48. Matsumoto T (1986) Kobunshi High Polymer Japan *35*: 675
49. May TC, Woods M (1978) 16th Ann. Proc. International Reliability Physics Symposium, p33
50. May TC (1979) Proc. Electronic Component Conference, p247
51. Geilhute M (1979) 18th IEEE Computer Soc. Int. Conf., p201
52. Fukuzawa I, Ishiguro S, Nanbu S (1985) 23th Ann. Proc. International Reliability Physics Symposium, p192
53. Kitayama A, Tabata H, Suzuki H (1986) The 4th International Microelectronics Conference, p462
54. Takada K (1987) NIKKEI NEW MATERIALS *25*: 25
55. Salmon ER (1987) Encapsulation of Electronic Devices and Components. Marcel Dekker Inc., New York
56. Nakagawa O, Sasaki I, Hamamura H, Banjo T (1984) J. Electronic Materials *13*: 231
57. Lee H, Neville K (1967) Hand book of Epoxy Resins. McGraw-Hill Inc.
58. Kamon T (1977) Kobunshi Ronbunshu *34*: 833
59. Gupta VB, Drzal LT, Lee CYC (1985) Polym. Eng. Sci. *25*: 812
60. Varma IK, Bhama PVS (1986) J. Composite Materials *20*: 410
61. Hasegawa K, Fukuda A, Tonogai S, Horiuchi H (1984) Kobunshi Ronbunshu *41*: 575
62. Day DR (1986) Polym. Eng. Sci. *26*: 362
63. Yamagishi M (1986) Electronic Engineering Japan *28*: 63
64. Flynn R, Cianciarulo AN (1968) SPE Journal *24*: 37
65. Bruins PF (ed) (1968) Epoxy Resin Technology. John Wiley and Sons Inc., New York
66. Potter WG (1970) Epoxy Resins. Arrowsmith J.W. Ltd.
67. Shechter LW (1956) J. Ind. and Eng. Chem. *48*: 86
68. Mih WC (1984) ACS Symp. Ser. No. *242*: 273
69. Shimbo M, Nakaya T (1986) J. Polym. Sci., Part B, Polym. Phys. *24*: 1931
70. Niino H, Noguchi S (1982) J. Appl. Polym. Sci. *24*: 2361
71. Son PN, Weber CD (1973) J. Appl. Polym. Sci. *17*: 1305
72. Smith JDB (1979) J. Appl. Polym. Sci. *23*: 1385
73. Smith JDB (1981) J. Appl. Polym. Sci. *26*: 979
74. Suzuki H, Sato M, Muroi T, Watanabe Y (1973) Preprint SPE 19th ANTEC p6
75. Hitachi Ltd. USP 3.859.379
76. Belani JG, Sporck CR (1978) SPE Tech. Pap. Annu. Tech. Conf. (Soc. Part Eng.) *24*: 557
77. Fujiwara H, Takahashi A., Fujii K, Nakamura K, Sekiya K, Suzuki K (1980) J. of Thermosetting Plastics Japan *1*: 14
78. Fujiwara H, Takahashi A, Fujii K, Nakamura K, Suzuki K, Yamada T, Horiuchi H, Tonogai S (1981) ibid. *2*: 1
79. Fujiwara H, Yoshinari T, Fujii K, Nakamura K (1983) ibid. *4*: 1
80. Shimbo M, Ochi M, Shigeta Y (1981) J. Appl. Polym. Sci. *26*: 2265
81. Baker E (1970) IEEE Trans. on Parts, Materials, and Packaging *PNP-6*: 121
82. Miyake K, Suzuki H, Yamamoto S (1985) IEEE Trans. Rel. *R-34*: 402
83. Schroen WH, Spencer JL, Bryanand JA, Clevelan RD, Metzgar TD, Edeards DR (1981) 17th Annu. Proc. Rel. Phys., p81
84. Isagawa M, Iwasaki Y, Sutoh T (1980) 18th Annu. Proc. Rel. Phys., p171
85. Kumata K, Iko K, Tabata H (1985) IEEE Trans. on Components, Hybrids, and Manufacturing Technology *CHMT-8*: 486
86. Nakamura Y, Tabata H, Suzuki H, Iko K (1986) J. Appl. Polym. Sci. *32*: 4865
87. Maxwell D, Young RJ, Kinloch AJ (1984) J. Mater. Sci. Lett. *3*: 9
88. Butta E, Levita G, Marchetti A, Lazzeri A (1986) Polym. Eng. Sci. *26*: 63
89. Bucknall CB, Partridge IK (1986) ibid. *26*: 54

90. Manzione LT, Gillham JK (1981) J. Appl. Polym. Sci. *26*: 889
91. So P, Broutman LJ (1982) Polym. Eng. Sci. *22*: 888
92. Kinloch AJ, Shaw SJ, Tod DA, Hunston DL (1983) Polymer *24*: 1341
93. Rosin P, Rammler E (1923) J. Inst. Fuel *7*: 29
94. Sundatrom DW, Lee YD (1972) J. Appl. Polym. Sci. *16*: 3159
95. Kanari K (1977) KOBUNSHI/High Polymers Japan *26*: 557
96. Pinheiro MDF, Rosenberg HM (1980) J. Polym. Sci., Polym. Phis. Ed. *18*: 217
97. Nishizawa J, Suyama S (1984) Polyfile *21*: 44
98. Nara S, Matsuyama K (1971) J. Macromol. Sci-Chem. *A-5*: 1205
99. Bremmer BJ (1964) I and EC Product Research and Development *3*: 55
100. Lum RM, Feinstein LG (1980) Proc. of the 30th Electronics Components Conference, p113
101. Feinstein LG (1980) ibid. p106
102. Thomas RE, Winchell V, James K, Schar T (1977) Proc. of the 27th Electronics Components Conference, p182
103. Pirron ED, Bobos GE (1977) J. of Electronic Materials *6*: 333
104. Sterman S, Marsden JG (1963) Modern Plastics *42*: 125
105. Plueddemann EP, Clark HA, Nelson LE, Hoffman KR (1962) Modern Plastics *39*: 135
106. Monte SJ, Sugerman G (1978) 33rd Annu. Tech. Conf. Section *2-C*: p1
107. Oliver JG (1969) J. Inorg. Nucl. Chem. *31*: 1609
108. Kusumoto Chemicals Ltd. (1984) Technical Bolletin *No. 201*
109. Hoechst Aktiengesellschaft Product Leaflet *8/84* (1984)
110. Cabot Corp.: Technical Report *S-34* (1984)
111. Ogata M, Kawata T, Kinjo N (1987) Koubunshi Ronbunshu *44*: 193
112. Sibayama K (1961) Koubunshi Kagaku *18*: 183
113. Tobolsky AV, Carlson DW, Indicator N (1961) J. Polym. Sci. *54*: 175
114. Shechter L, Wynstra J (1956) Ind. Eng. Chem. *48*: 86
115. Kakiuchi H (1987) Polyfile *24*: 2
116. Williams ML, Landel RF, Ferry JD (1955) J. Am. Chem. Soc. *77*: 3701
117. Funaki Y, Yamaguchi K (1987) NIKKEI MICRODEVICES *22*: 55
118. Onuki J, Koizumi M, Araki I, In press
119. Mooney M (1951) J. Colloid Sci. *6*: 162
120. Doi H, Tsuchimoto M, Ono M (1985) Denki Kagaku *53*: 282
121. Ikoh K, Tabata H (1986) Zairyo Gijyutsu *4*: 27
122. Asai S (1986) Proc. of the IEEE *74*: 1623
123. Yamaguchi K, Igarashi K, Tabata H, Suzuki H (1986) IEEE Trans. on Components, Hybrids, and Manufacturing Technology *CHMT-9*: 370
124. Sundstrom DW, Walters LA, Goff CS (1969) Modern Plastics *10*: 101
125. White RP jr (1974) Polym. Eng. Sci. *14*: 50
126. Roller MB (1975) ibid. *15*: 406
127. Boyer E, Macosco CW (1976) AIChE Journal *22*: 268
128. Lipshitz SD, Macosco CW (1976) Polym. Eng. Sci. *16*: 803
129. Manzione LT (1981) ibid. *21*: 1234
130. Rojas AJ, Adabbo HE, Williams RJJ (1981) ibid. *21*: 634
131. Saeki J, Kaneda A, Tanaka M, Tsuchiya K (1984) J. of 9th International Congress on Rheology, p679
132. Kaneda A, Aoki M, Saeki J (1977) Proc. of 1st International Congress of Polymer Processing, p344

Editor: K. Dušek
Received October 5, 1987

Synthesis and Structure of Macromolecular Topological Compounds

Yu. S. Lipatov, T. E. Lipatova, and L. F. Kosyanchuk
Institute of Macromolecular Chemistry, Academy of Sciences of the Ukrainian SSR, Kiev, USSR

The present review discusses, systematizes, and critically summarizes the results of investigations in the field of synthesis of macromolecular topological compounds (polycatenanes and polyrotaxanes). Since the chemistry of such polymers can be understood only on estimating the results of synthesis and investigations of their low-molecular-weight analogs, the review also systematizes recent advances in this field. Considerable attention in the review is paid to issues of synthesis and investigation of polyrotaxanes. These studies are concerned with the statistical synthesis (by Agam and Zilkha), directed synthesis (by Maciejewski), and synthesis with elements of directionality.

1 Introduction 50
2 Catenanes, Rotaxanes: Nomenclature 51
3 Natural and Synthetic Macromolecular Topological Compounds
  and Their Features 56
4 Polyrotaxane: Synthesis Techniques 59
5 Synthesis of Polyrotaxanes with Elements of Directionality 65
  5.1 Ordering of Macrocycles by Means of Complexing 65
  5.2 Synthesis of Polyrotaxanes with the Use of Swarm Complexes and
      Investigation of Their Structure 69
6 Conclusion 73
7 References 73

1 Introduction

The number of publications dealing with compounds of a new unusual type, where molecules are bound together by mechanical entanglement without the participation of a chemical bond, has been increasing in recent years. Such compounds include catenanes, rotaxanes, and knots, which can be schematically represented as follows [1-6]:

Catenane   Rotaxane   Knot

Publications describing the synthesis of these compounds first appeared more than 30 years ago, and not much later reports of the discovery of similar substances in nature also appeared. The data on the synthesis of catenanes, rotaxanes, and knots were, in early 1970s, summarized and systematized in a monograph by Schill [6]. Many new results have recently been published [7-16], which considerably supplement and expand the concepts presented in the monograph. A new trend involving the synthesis of macromolecular analogs of the compounds under consideration has emerged, but the number of studies conducted along this line is as yet not numerous [17-23].

The purpose of the present review is to systematize and critically summarize the recent advances in the chemistry of such polymers.

All authors describing one or another compound with a nonchemical bond often employ the terms "topological compound" and "topological bond" [17,24-27]. These concepts should be briefly dealt with in order to explain the meaning attributed to them.

Consider two chemically unbound molecules: they can continuously and randomly change their location with respect to each other. There exists, however, such a relative arrangement of them when one of the molecules cannot change its position with respect to the other unless its chemical bond is broken. In this case, a system of two or more molecules becomes an individual compound whose integrity is due to a new type of bond, a topological one, the strength of the bond being determined by the weakest chemical bond in the components of the compound, and for rotaxanes, also by the size of bulky substituents with respect to the size of the macrocycle interior [2,28].

Compounds which contain topological bonds are called topological [29]. They represent topological isomers of two separate molecules of the components [1] (two rings for a catenane; a ring and a linear molecule with large bulky substituents, "dumb-bells", for a rotaxane [6,21,29]).

There exists one more type of topological isomerism; a particular case of such a type, "a molecule in a cage" [26] is of interest. Formally into this category may be placed inclusion compounds (clathrates) of urea and thiourea, cyclodextrins with various substances [30,31], cryptates [32], carbidic organometallic clusters [33,34], inclusion compounds of graphite [35] and of other laminar compounds [36]. The forces retaining the "guests" (molecules, atoms, ions) in internal voids of crystals of host molecules are of a different nature. This may be a Van der Waals or various types of donor-acceptor [36], or electrostatic [37] interactions. Consequently, the stability of such compounds as a whole will differ as well: they either dissociate easily due to the destruction of intermolecular bonds with a disturbance of the crystalline state [38,39] or exhibit high stability [40].

Theoretical approaches to the synthesis of compounds constructed according to the "a molecule in a cage" principle are discussed in [41]. These compounds represent a "guest" molecule or molecules, trapped by a host molecule with an enclosed structure similar to an egg shell. The author of Ref. [41] believes that enclosed shell-like compounds can be produced either by completing the geometry of a "pre-shell" molecule to a spatially enclosed one or by forming a cascade of branched oligomers and polymers. In the former case, a simple shell molecule is formed, the synthesis of which in the presence of a guest molecule makes it possible to obtain lowmolecular topological compounds schematically represented as follows [41]:

As pre-shell molecules cyclodextrins [41] and trichinocene-like compounds can be used [26]. The author of Ref. [41] offers the scheme of such a closing of a cyclodextrin molecule into a "shell" around a guest molecule in its interior:

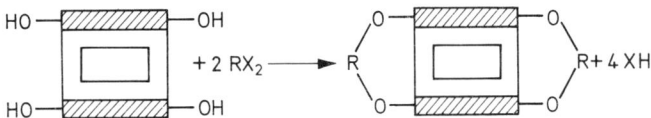

In the opinion of Maciejewski [41], the same compounds can be obtained by polycondensation of monomer XRY (X, Y are reaction groups capable of interaction with each other, such as OH and COOH) in a good solvent. Here some of the solvent molecules, in the course of branching, are introduced inside a spherical shell of the cascade of branched molecules. He shows this schematically as follows:

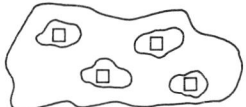

Such a formation, however, can hardly be identified as an individual compound. The stability of the system under consideration will be determined by thermodynamic parameters of the polymer-solvent interaction. Under certain conditions the system can undergo phase separation right up to pure phase, i.e. a complete removal of the solvent from the bulk of the polymer without destruction of the latter will occur [42].

2 Catenanes, Rotaxanes: Nomenclature

Since the chemistry of polyrotaxanes and polycatenanes can be well understood only on estimating the results of synthesis and investigation of their low-molecular analogs, we regarded it reasonable to look at the publications concerned with the compounds.

The existence and possibility of synthesis of catenanes and rotaxanes were suggested as early as the beginning of this century by Wilschetter long before the first large cycles were obtained [43]. It was, however, not until 1964 that a catenane was first

obtained and identified by Schill et al. [44]. The results of further successful studies on the creation of catenanes were systematized by Schill in a monograph [6], the historically first review publication dealing with such studies. An analysis of the available experimental data allowed Schill to distinguish the following ways for the synthesis of topological compounds: a directed synthesis [5,8,9,11,12,45-47]; a statistical synthesis (for catenanes and rotaxanes) [9,13,2,48]; and a synthesis based on the Mobius strip principle (for catenanes and knots) [3,4,7,9,14,49].

The directed method of synthesis of catenanes and rotaxanes was developed by Schill with co-workers, who are at present continuing research in this field [7,10-13,47]. Since the approaches to and requirements for such syntheses are well scribed in a monograph [6], we do not considered it necessary to dwell on it in detail. The essence of such methods consists in, (1) preparing so-called "precatenanes" ("prerotaxanes") and, (2) breaking the "temporary" chemical bonds between the molecular components with the aim of isolating the desired final product. A prerotaxane (precatenane) is an intermediate compound, when a linear (cyclic) molecule is connected by 2–3 chemical bonds to a macrocycle and disposed across the ring plane. The synthesis of such compounds is very complex and laborious; it is accomplished by conducting several series of multistage transformations.

The statistical method of synthesis of topological compounds can be divided into a truly statistical synthesis and a synthesis with elements of directionality. Preparing catenanes by the statistical method is described in detail by Schill in his monograph [6]. The statistical technique of synthesis of rotaxanes will be presented here in greater detail. The scheme of the technique is as follows: bulky end groups are connected to a linear molecule in the presence of a macrocycle. Such a reaction, where the macrocycle remains unchanged, results in the formation of some amount of rotaxane and a molecular "dumb-bell":

Similar results can be expected with cyclization of a chain in the presence of a "dumb-bell" compound [50] or with dissociation of the latter, followed by regeneration in the presence of macrocyclic compounds [2].

. The synthesis of rotaxanes after such a scheme has been studied in detail by Harrison [51]. By binding a macrocycle of the form with chloromethylated styrene-divinyl benzene copolymer and then by subjecting the system to treatment by decane diol-1,10

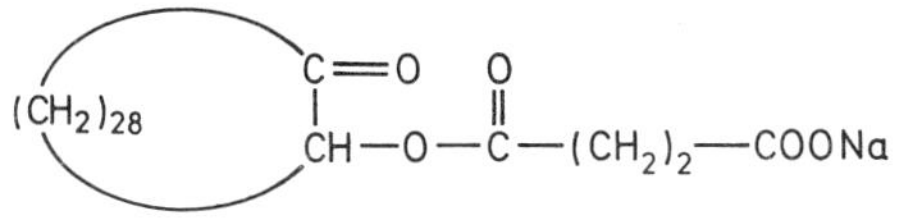

$$(CH_2)_{28} \quad \begin{matrix} C=O & O \\ | & \| \\ CH-O-C-(CH_2)_2-COONa \end{matrix}$$

and triphenylchloromethane in a pyridine-toluene solvent mixture, repeated 70 times and followed by hydrolysis by sodium carbonate in boiling methanol, he succeeded in obtaining a mixture which contained 6% rotaxane:

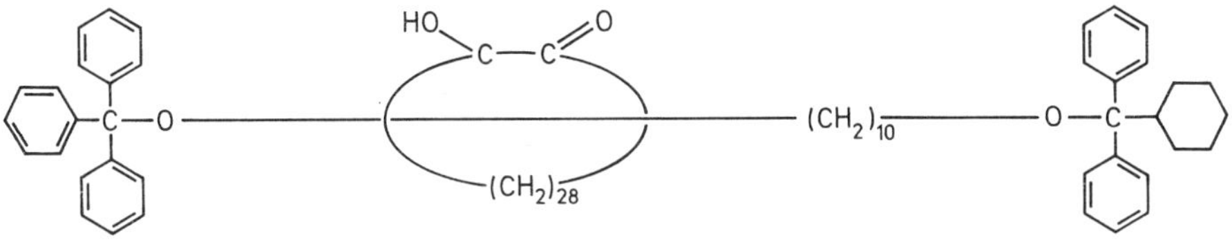

[Structure showing trityl-O-(CH_2)_10-O-trityl dumb-bell threaded through a macrocycle containing HO-C-C=O and -(CH_2)_{28}- groups]

The product was isolated by a chromatographic technique as an oil. It turned out to be stable on heating up to 200 °C. The IR spectrum of the product corresponded to that of an equimolar mixture of the macrocycle and "dumb-bell". Thin-layer chromatogram of the sample indicated an absence of unbound molecular components.

The authors of Ref. [50] attempted to synthesize a rotaxane by cyclization of diethyl ether of hexacosanedionic acid-1,26 or diethyl ether of pentacosanedionic acid-1,25 in a "dumb-bell" solvent (1,6-bis[diisopropylmethoxy]-hexane; 1,6-bis[dicyclohexylamino]-hexane; 1,10-bis[dicyclohexylamino]-decane) by mixing a sodium suspension in the solvent with a solution of the above-mentioned acids. This attempt failed, however.

Harrison [2] later described a statistical synthesis of a rotaxane apparently due to a partial detachment of side groups of a "dumb-bell" under the action of some compounds and their subsequent reconnection in the presence of a macrocycle. According to the laws of statistics, some of the cyclic molecules are threaded on such molecular fragments which exist in an equilibrium with the "dumb-bell". Subsequent regeneration of the "dumb-bell" results in the formation of rotaxanes. Used as the dissociating "dumb-bell" molecules were 1,3-di[tri-4-tertbutylphenylmethoxy]-tridecane; bis-dicyclohexyl acetate of decane diol-1,10. The detachment of end groups occurs in the presence of β-napthalenesulfonic acid (220 °C, 1 min). Harrison proved the formation of rotaxanes by the comparison of 1 H-NMR-spectra of the compound obtained and an equimolar mixture of the corresponding macrocycle and "dumb-bell" solvent as well as by the formation of the initial macrocycle on dissociation of the rotaxane.

As to the identification of catenanes and rotaxanes, this is hard to attain by relying on any single method. The difficulty is that rotaxanes and catenanes in no way differ from the mixture of their individual components. Such compounds can only sometimes be identified by mass spectrometry, as predicted by Kostjanovskij [25] and demonstrated by Schill [6,46]. If this method fails to result in unambigous identification, other research techniques (chromatography, IR and NMR spectroscopy, X-ray structure analysis, electron microscopy) have to be applied. Only comparing the results of independent techniques makes it possible to identify the products correctly and to draw definite conclusions on the structure of compounds obtained.

The development of methods for the synthesis of topological compounds with elements of directionality began as early as studies by Schill [6]. Thus, the synthesis by this method involves an attempt to bind temporarily the chains with the macrocycle, which results in compound 1 with a prerotaxane structure (see the diagram below) and its isomer 2, which are in conformational equilibrium [6]. After connection of bulky end groups and breaking of "temporary" chemical bonds, a mixture is formed consisting of a rotaxane, a "dumb-bell", and a macrocycle:

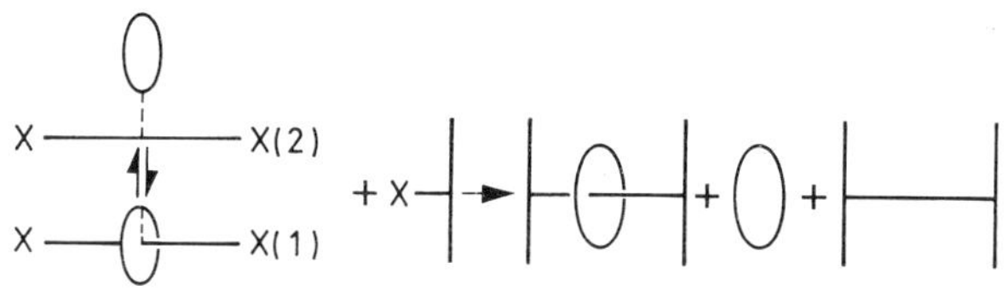

Obtaining rotaxanes in this way, as well as the directed synthesis, is laborious and is accomplished by a number of complex syntheses-transformations. It should be emphasized that — due to a very insignificant yield of the desired products in the statistical synthesis of topological compounds and the great laboriousness of the directed method (which calls for a great ingenuity and a high experimental skill) — it is very difficult to give preference to any one of the methods for producing such substances. This fact encourages search into other approaches to catenane and rotaxane synthesis. The methods for obtaining them have been considerably expanded and simplified recently. In particular, ways for a synthesis with elements of directionality, where the properties of complex compounds are employed, have been described [52–54].

Great interest has been shown in a series of reports by French researchers [52,55–60] published between 1984 and 1986 and describing the synthesis [52,55,58–60] and investigation into the structure [56,57] of metal-catenanes:

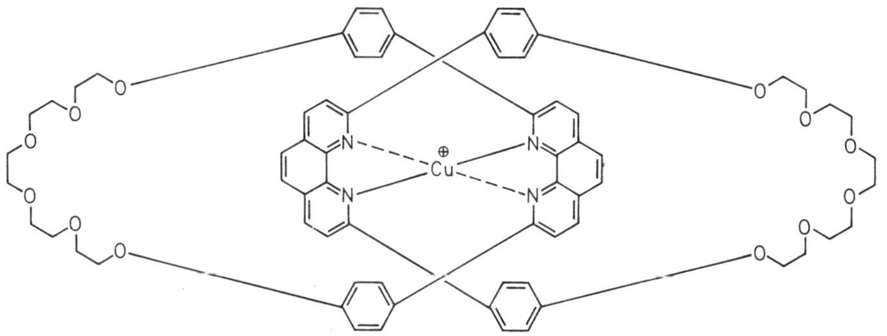

This approach to the synthesis relies on the capability of an ion of univalent copper to form strong complexes both with two linear molecules the macrocycles of which can be produced by cyclization with one such molecule and a already formed cycle. In this case, Cu^+ orients them spatially so that closing of linear molecules into a ring gives rise to a metal-catenane:

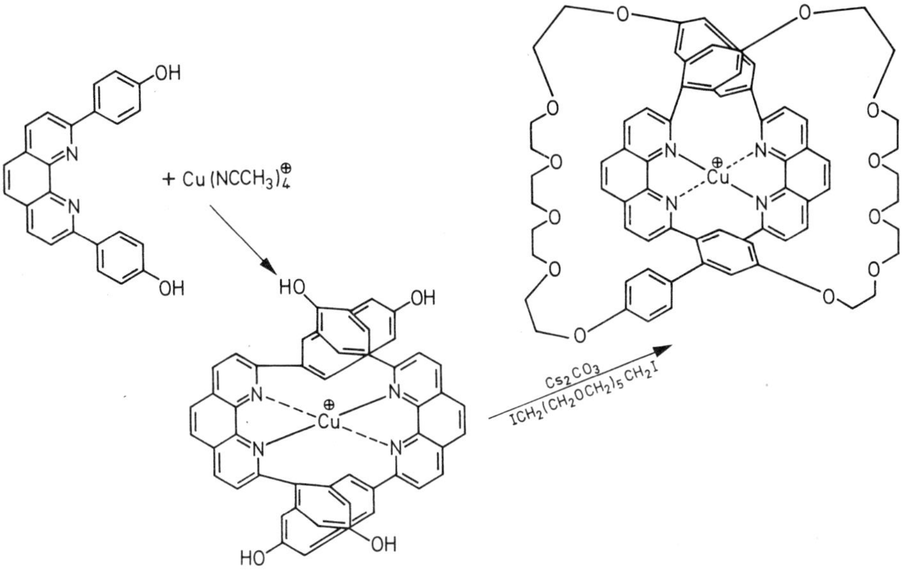

An electrochemical demetallation shows an exceptionally high stability of the metal complexes being formed and results in the formation of a catenane tetradentate ligand.

The identification of the substance was based on ^{1}H-NMR spectra and mass spectrometry. The yield of desired products amounted to 19%, the number of stages in the process being as few as three.

Of no less interest is a series of studies on the production of rotaxanes conducted by Japanese researchers, where large voluminous Co^{3+} complexes were first used as the end groups and α- and β-cyclodextrins as the macrocycles. The synthesized rotaxanes are represented by compounds

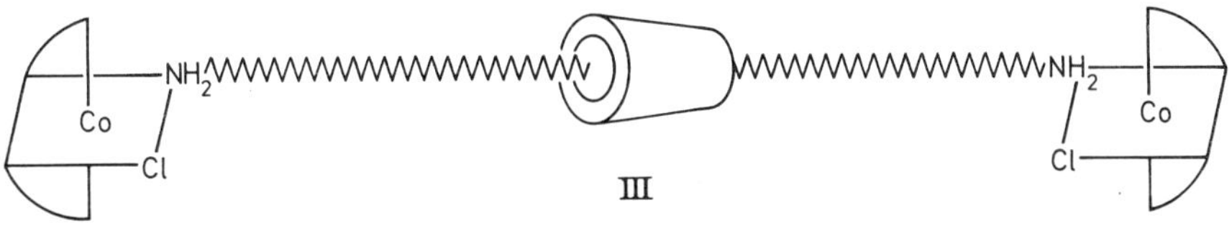

Refs. [53, 61)]

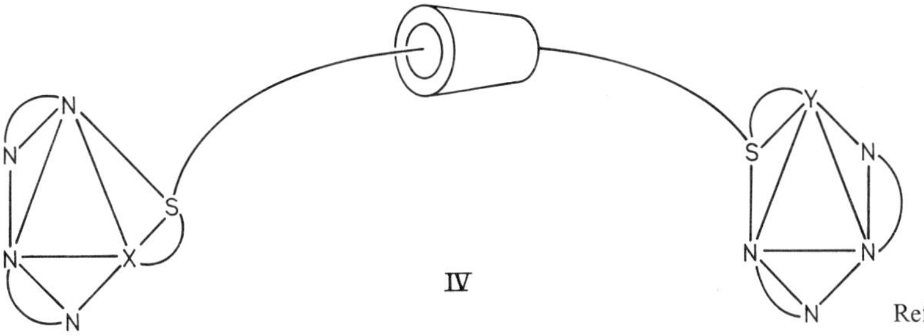

Refs. [54, 62)]

The approaches to obtaining these rotaxanes are similar to the procedure of classical statistical syntheses where a "dumb-bell" is formed in the presence of cycles or another identical bulky substituent is connected to a linear molecule having one such bulky substituent in the presence of cycles. Elements of directionality are introduced to this synthesis by the complexing ability of cyclodextrins. A tendency of cyclodextrins, which are compounds consisting of glucose residuals interconnected by 0-glucoside bonds [63], to form inclusion compounds with various substances has long been known [21,64–69]. Utilizing this property of cyclodextrins, Lüttringhaus et al., as early as 1958, attempted to obtain catenanes by cyclization of long-chain dithiols in the form of α-cyclodextrin inclusion compounds [70,71], but the attempt failed. They intended to perform the cyclization by oxidation of dithiols to sulfides under the action of air in the presence of Cu^{2+}. The failure was attributed to the fact that dithiol was not oxidized because of shielding of side chains by cyclodextrin molecules. The use of cyclodextrins by Japanese investigators resulted in the formation of rotaxanes. The yield of products with the use of α-cyclodextrin was higher due to a better matching of the cycle size to the size of the penetrating molecule and the size of the end groups.

The authors of Refs. [53,54] proved the rotaxane formation by the UV, C^{13}-NMR, and circular dichroism spectra. Electronic spectra of rotaxanes III and IV are typical for compounds with chromofores (CoN_6Cl) [53,61] and $[CoN_6S]$ [54,62], respectively. The similarity of signals of NMR spectra of an individual cyclodextrin and rotaxane rules out the possibility of chemical interaction between cyclodextrin and metal-containing fragments. The spectral pattern of bands characteristic for shifts of the binuclear complex in rotaxane are similar to the spectrum of the individual component but more complex. The authors of Ref. [61] attribute this phenomenon to the formation of rotaxane structures. Since α-cyclodextrin is chiral and has the shape of a truncated cone, the symmetry of signals of the "dumb-bell" components with bulky metal complexes at the ends is disturbed, and broadening of signals in the spectrum occurs. A weak dichroic absorption band $\Delta\varepsilon = +0.04$ is observed in the circular dichroism spectrum of rotaxane III; it has been attributed to induction of an optical activity in the complex fragment. The formation of a rotaxane is also indicated by an absence of a dichroic absorption in this region for a mixture of cyclodextrin and "dumb-bell" component. The rotaxane formation is confirmed by elementary analysis and by separation of an unchanged binuclear complex of cyclodextrin.

To conclude, let us briefly mention the nomenclature of topological compounds. It is generally accepted at present to call such compounds according to the nomenclature proposed by Schill [6]. A generalized theory of description and illustration of topological compounds is presented in Ref. [72].

3 Natural and Synthetic Macromolecular Topological Compounds and Their Features

Rotaxanes, catenanes, and knots of the nucleic class were discovered in nature soon after the synthesis of the first topological compound. In 1967, Vinograd et al. [73–75] discovered and Wang [76] isolated a catenane of DNA. Using electron spectroscopy, the presence of interlocked cyclic double-helix mitochondrial DNA in Hela cells as well as in leucocytes of people ill with leukemia was demonstrated. An increased content of catenane DNA was later found in people with various forms of cancer [77], as was rotaxane DNA [78]. Soviet scientists discovered a cyclic RNA [79]; by electron microscopy it has been concluded that cyclic RNA exists in the form of a knot. Such forms of existence of nucleic acids continue to interest biologists and chemists, which is evidenced by recent studies [80-82].

A possible mechanism of the biogenesis of catenane DNA, based on the Mobius band properties, was proposed by Schill and Zürcher [83]. When a Mobius band, twisted n times and its ends glued together, is cut along the centreline, catenanes with various degrees of interlocking are obtained at even n, and various knots, at an odd n (see Fig. 1) [6]. The possibility of constructing cyclical DNA molecules on the principle of the Mobius band was suggested long ago by Wasserman [84] and, as demonstrated in Ref. [85], does not contradict the properties of cyclic DNA. Replication of a two-thread Mobius DNA will be equivalent to cutting, which may ultimately result in the formation of catenane and knot nucleic acids.

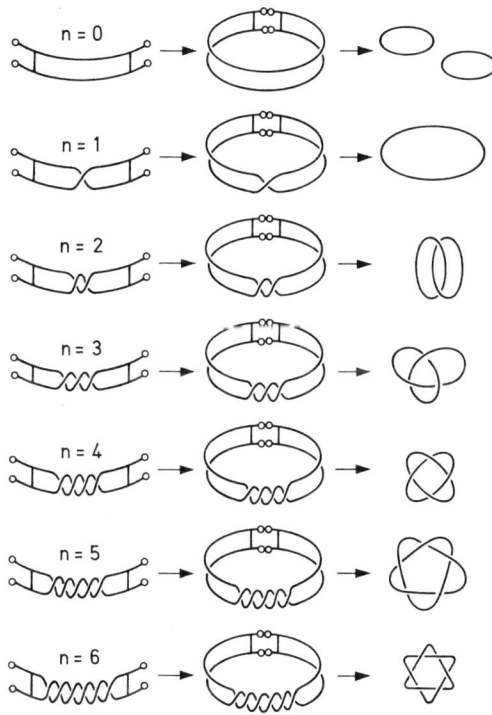

Fig. 1. Structures forming at cycling of two-strip bands with number of half-turns n after breaking of transverse bonds

The above-described compounds may already be considered as natural topological polymers. Synthetic compounds of a similar type (polycatenanes, polyrotaxanes) have as well been described in the literature; however, not all the results may be regarded as correct. Thus, the author of Ref. [21] tries to treat the system consisting of a sparsely cross-linked polymer network and a linear polymer with large bulky substituents as a system having a polyrotaxane structure. Also Frisch et al. represented various interpenetrating polymer networks as compounds with a polycatenane structure [86-90]. Investigations by Lipatov et al. [91] established, however, that such materials are characterized by a microheterogeneous distribution of components and hence there is no mutual penetration at the molecular level, i.e. the formation of rotaxane and catenane structures in such systems is unlikely. The report by Callahan and Frisch [20] of obtaining a polycatenane based on cross-linked polybutadiene and cyclic peptide (valinomycin) is also unconvincing for the same reason.

An interesting possibility for the formation of linear polycatenanes is the statistical synthesis technique proposed by Greek researchers [92-98]. The technique consists in cyclization of a bifunctional long-chain compound with another bifunctional compound (for example, a reaction between a dibasic acid chloride and diamine) at the phase interface in the presence of large cycles (crown-60, oligomeric cyclosiloxanes) whose interior allows two monomer molecules to penetrate into the ring simultaneously. The macrocycles form a monomolecular layer at the phase interface. The possible cyclization due to the interaction of an acid chloride molecule with a diamine or diol molecule can result in the formation of true polycatenanes, which is the case when two formed cycles penetrate a cycle of another nature already present in the system. Unfortunately, the last report was in 1976 and no more reports have been published.

The problems of synthesis and investigation of polyrotaxanes are presented in the literature by a number of studies. These are researches in the field of statistical synthesis [99,100], directed synthesis [100-102], and synthesis with elements of directionality [103], which have been pursued by three groups of researchers. Before discussing these studies, specific features of the structure of polyrotaxanes should be explained. The principal difference of these polymers from their low-molecular analogs is that long polymer chains are present in the system instead of a small linear molecule on which the macrocycles are "threaded". Also, no large bulky substituents or end groups are needed for such compounds [100], and, therefore, in polyrotaxanes described in the literature these groups are either absent [100,104] or their dimensions allow the cycles to "glide off" and "thread on" the polymer chain [99].

The stability of the polyrotaxane structure can result from both the intermolecular interaction (of the type of hydrogen, donor-acceptor, and Van der Waals bonds) between the macrocycles and polymer [105] and a conformation change due to mobility of their individual fragments (e.g., a change of a cylindrical conformation to a conical one is possible owing to vibratory motions of residuals in cyclodextrins [104]). A decrease in the diameter of one of the truncated cone bases enhances the stability of polyrotaxanes. Moreover, polyrotaxanes are stabilized by the very nature of the polymer chain (the possibility of having a set of conformations) [106]. With a large length of the polymer chain, cyclic molecules can "glide off" only at the polymer molecule ends, and this effect is minimized by the solid state of polyrotaxanes. Thus, the number of rings per one linear chain in a polyrotaxane formed from dibenzo-58-crown-19 and polyethylene glycol PEG-400 with triphenylol end groups remains unchanged after 60 days storage at room temperature [99]. The stability of a polyrotaxane formed from β-cyclodextrin and dehydrochlorinated poly(vinylidene chloride) [104], is due to the conformation of the polymer chain after a treatment with NaOH inhibiting the movement of rings along the chain:

To conclude, the nomenclature of macromolecular topological compounds should be briefly mentioned. In fact, it does not exist. The use of the Schill's nomenclature is difficult, since the name of polymeric catenanes and rotaxanes, where the chain consists of many interlocked rings or rings threaded on a long macromolecule, becomes cumbersome and hard to pronounce. Therefore, trivial names are often used.

4 Polyrotaxanes: Synthesis Techniques

The basic regularities of the statistical synthesis of polyrotaxanes have been established by Agam and Zilkha [99]. They obtained polyrotaxanes by a statistical threading of crown polyesters on polyethylene glycols. Polyethylene glycols with a molecular mass of 400, 600, and 1000 were selected as linear chains, and crown polyesters with different numbers of atoms in the cycle (30, 44, and 58), as macrocycles. The reactants were mixed, heated, and an equilibrium mixture was "frozen" by the addition of naphtilene-1,5-diizocyanate with the formation of polyurethane. The number of bound cycles was determined after a selective separation on silica gel.

A chemical interaction (or the formation of a strong complex) between polyurethane and macrocycle in the isolated final product was ruled out by additional studies. Thus, adsorption of premixed crown polyesters and polyurethane on silica gel resulted in a complete extraction of cycles (96%), which demonstrated an absence of strong complexes. Dissolution of the isolated final product of reaction in dimethylformamide (DMF) at various concentrations yielded no crown polyesters, i.e. no trapping of macrocycles by polyurethane took place. Detection of a considerable amount of a free cycle in such solutions after 60 days showed an absence of chemical interaction between polyurethane and the cycles. Based on the data obtained, the authors of Ref. [99] identified the product as a polyrotaxane.

The effect of the cycles-to-chains molecular ratio, chain length, cycle dimensions, their concentration in the initial system, and reaction temperature on the product yield has been studied. It has been shown that the maximum penetration of cycles by a chain is attained at an equimolar ratio of the components (a macrocycle consisting of 58 atoms and PEG-600). It has also been established that a dilution of the system, e.g. by an inert solvent, i.e. a decrease of the concentration of components, reduces the yield of polyrotaxanes. Changing the temperature in the process of polyrotaxane formation exerted no effect on the desired product yield [99].

Employing the basic laws of statistics and taking into account the factor of the required orientation of cycles with respect to chains, the size of the ring, the length of the chain, and their ratio affecting the dynamics of the "threading", an equation was suggested [99] describing the dependence of the threading degree N

$$N = K \frac{m_c m_g (1 - e^{-n_c/\pi n_g} \cdot n_c n_g^\beta \theta)}{\vartheta} \tag{1}$$

where N is the number of penetrations; $m_c m_g$ are the numbers of rings and chains respectively; n_c, n_g are the numbers of atoms in a cycle and a chain; ϑ is the overall volume; θ is the penetration angle, depending on the ring radius (r) and chain diameter (d) and determined as $\cos \theta = \frac{d}{2r}$. The calculation of is based on the assumption that the ring is fully stretched and on the use of tabular values for bonds and Van der Waals radii; K is a constant; and β is a constant characterizing the degree of curling-up of chain molecules, which affects the "threading-on" and "gliding-off" of rings ($1 < \beta < 2$). The values were selected as: $K = 0.195$ and $\beta = 1.3$. The parameters of the initial system, obtained with this equation, show good agreement with experimental data. It should be noted, however, that the use of Eq. (1) in an unchanged form for other systems (e.g. with other cycles and penetrating chains) is difficult since, apart from an empirical selection of the K and β values, corrections for the change of the shape of rings should be made in the equation, and this will greatly complicate such calculations. Thus, an attempt to apply Eq. (1) to their systems by the authors of [61] resulted in understated values of the threading degree (N) as against experimental data.

The optimum conditions for the penetration, found in [99], were employed in the synthesis of a polyrotaxane based on PEG-600 and dibenzo-59crown-19 as well as a polyrotaxane obtained by an anionic polymerization of oxyethylene under the action of potassium polyethylene glycolate in the presence of a macrocycle. A polyrotaxane-structure compound was identified with the aid of spectral techniques and chromatography. An absence of a chemical interaction between the ring and chain was proven by hydrolysis of blocking groups, which resulted in triphenyl methanol, a linear chain of polyoxyethylene glycol, and an unchanged macrocycle. The polyrotaxanes obtained, occupy an intermediate position between their components with regard to their solubility — they dissolve in hot cyclohexane, a good solvent for the macrocycle and a poor one for linear glycol. In addition, invariability of the NMR spectra of the synthesized polyrotaxanes over a temperature range of $+25$ to -30 °C has been shown, attributed by the authors of [99] to the possibility of greater freedom of movement of components in the polyrotaxane.

The same publication presents interesting results on thermal stability of the polyrotaxanes. For an existing equilibrium [28,51],

$$\text{Rotaxane} \underset{k'}{\overset{k}{\rightleftarrows}} \text{Ring} + \text{Chain},$$

the rotaxane equilibrium constant (K), dissociation rate constant (k), and formation rate constant (k') have been determined:

Temperature, °C	K, mole/l	$k \times 10^4$, s^{-1}	$k' \times 10^4$, s^{-1}
130	0.133	0.112	0.84
150	0.282	0.206	0.73
170	0.477	0.485	1.02

Studying the decomposition of 5% solutions of polyrotaxanes in various solvents (protogenic octanol-2, nonpolar xylene, and diglyme which is close in nature to the rotaxane components) demonstrated that the rate of polyrotaxane dissociation in diglyme is the same as in bulk ($k_{150} = 0.189 \times 10^{-4}$ s^{-1}); in octanol the dissociation rate increases ($k_{150} = 0.53 \times 10^{-4}$ s^{-1}); and in xylene it drops ($k_{143} = 0.059 \times 10^{-4}$ s^{-1}). The authors of Ref. [99] explained the difference in the dissociation rate as being due to interaction of polyrotaxane components both with one another and with solvent molecules. The presence of diglyme brings about no changes in the interaction between the crown polyester and linear glycol and the dissociation rate remains unchanged, but protogenic octanol-2 forms hydrogen bonds with rotaxane components and thereby increases the rotaxane dissociation rate. Xylene, as a nonpolar solvent, favors strengthening the ring-chain interaction and, being hydrophobic, promotes a compression of a molecule, which reduces the disintegration rate.

The stability of polyrotaxanes was accounted for by the macrocycle tendency to assume such a conformation which would assure the maximum dipole-dipole interaction between oxyethylene links of the chain and ring. This was confirmed by kinetic studies which demonstrated the polyrotaxane dissociation activation energy to amount to 15.9 kcal/mole and the ring-on-chain threading i.e. polyrotaxane formation, activation, energy to 3.4 kcal/mole.

Obtaining polyrotaxanes by the statistical method through polymerization of various monomers in the presence of macrocycles was attempted by Maciejewski [100], who investigated the thermal radical copolymerization of vinylidene chloride, methyl methacrylate (MMA), acrylonitrile (AN), acrylamide, and their mixtures with acetonaphtylene in aqueous and dimethylformamide solutions in the presence of β-cyclodextrin or of its acylic derivative. All these attempts failed, however. Only, when copolymerization was initiated by γ-radiation with an exposure dose of 5 Mrad, did some of the above-listed monomers form polymers exhibiting optical activity. However, the rotation of the initial β-cyclodextrin was 162°, while that of the obtained products only 1°–6°. Such a small polarization plane rotation angle could result only from an insignificant amount of macrocycles in the synthesized compounds. Therefore, for the purpose of a more successful synthesis of cyclodextrin-based rotaxanes, a number of researchers [100,101] tried to introduce elements of "directionality" into their synthesis by means of creating a certain ordering of the initial polymerizing mixture. They also proceeded in their studies from the concepts of the tendency of cyclodextrins to the formation of inclusion compounds with various substances and of the possibility of such compounds having structures in the form of "channel" complexes [21,64–69]. Using the results of these studies, Maciejewski et al. [100,102, 107–109] as well as Ogata et al. [101] assumed that the adducts of β-cyclodextrin with various monomers (inclusion compounds) have the form of a channel of β-cyclodextrin molecules, which contains monomer molecules in its inner spaces. It should be pointed out that isolated inclusion compounds of β-cyclodextrin with guest molecules capable of both the polymerization [100,102] and the polycondensation [101] reaction were used for the synthesis of polyrotaxanes. Thus, polyamides [101] containing an included β-cyclodextrin were obtained by conducting polycondensation at the phase interface between inclusion compounds of β-cyclodextrin with various diamines (hexamethylenediamine, *p*-xylylenediamine, *m*-xylylenediamine) and dibasic acid chlorides. IR spectra of such polymers contain absorption bands characteristic of amides and cyclodextrin, and the elemental composition of the synthesized compounds are in a good agreement with the calculated one. Unfortunately, the researchers failed to clearly identify the products.

Subjecting inclusion compounds, isolated beforehand, to γ-irradiation, Maciejewski et al. [21,100,104] managed to obtain stable polymeric adducts with a polyrotaxane structure only when adducts of β-cyclodextrin with vinylidene chloride or allyl chloride served as initial adducts. Polymerization of the rest of the monomeric adducts resulted in unstable compounds, dissociating into a homopolymer and the initial β-cyclodextrin in hot water.

These researchers attribute the stability of poly(vinylidene chloride), containing a great amount of β-cyclodextrin, both to the formation of hydrogen bonds between molecules of cyclodextrin which define the channel, and to weak H-bonds between hydroxyl groups of β-cyclodextrin and chlorine atoms of poly(vinylidene chloride) [21,105]. Interaction of β-cyclodextrin molecules with one another [105], can give rise to seven H-bonds between OH-groups of two adjacent cyclical molecules on one side of cyclodextrin and seven on the other side, the poly(vinylidene chloride) chain close against the inside walls of the channel, thereby impeding vibratory motions of β-cyclodextrin. Such an arrangement of the polymer in the β-cyclodextrin channel makes it impossible to break only one or two H-bonds, which is inherent in low-molecular adducts of cyclodextrins. Separating a β-cyclodextrin molecule from the polymeric adduct requires breaking the seven hydrogen bonds simultaneously, which will call for a much greater energy expenditure, i.e. the polymeric adduct is more stable. On the other hand, the formation of H-bonds between β-cyclodextrin and the polymer also stabilizes the final polymer products. It has been shown that the longer the chain, the more such bonds will be formed and the more stable will be the resulting compound. If, however, such an explanation of stability of the products is assumed, then it is incomprehensible, why no stable polyrotaxanes result from polymerization of monomeric adducts of β-cyclodextrin with MMA which has a carbonyl atom of oxygen and is capable of forming stronger hydrogen bonds than does Cl ... OH. However, one can assume that the inside surface of a β-cyclodextrin molecule is hydrophobic [66], which should promote the stability of the product formed by polymerization of a styrene adduct and β-cyclodextrine. However, it is not the case. The exceptional behavior of the adduct of β-cyclodextrin with vinylidene chloride is ascribed [21,104], in the case of polymeric adducts of β-cyclodextrin with polystyrene, poly(methyl methacrylate) (PMMA), polyacrylonitrile (PAN) to sufficient free space remaining in the channel and the mobility of β-cyclodextrin segments being considerable, which results in instability of polymeric adducts. This explanation, however, is not altogether convincing, since styrene and MMA molecules are larger than vinylidene chloride ones. We believe the instability of the products of polymerization of inclusion compounds of β-cyclodextrin with MMA and styrene to stem from the large size of side substituents at the double bond of monomers. Vinylidene chloride is freely accommodated in the interior and, there can be three such molecules [105], while styrene and MMA molecules can be arranged in the β-cyclodextrin interior only so that atoms at the double bond protrude from the cycle interior (see Fig. 2). As a result of polymerization of such adducts, β-cyclodextrin turns out to be "put" on the side substituents and easily "glides off" from them after "hard" treatment of the synthesis products.

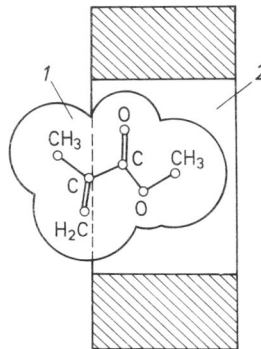

Fig. 2. Inclusion compound of β-cyclodextrin with methyl methacrylate (*1* — methyl methacrylate; *2* — cross-section of β-cyclodextrin)

Based on the experimental data on vinylidene chloride polymerization in the presence of linear dextrin, a chemical interaction between poly(vinylidene chloride) and β-cyclodextrin was ruled out [102]. The results of the experiments demonstrated an absence of a chain transfer to linear dextrin, i.e. an absence of chemical bonds. In addition, comparing the data of investigation of the structure of obtained compounds [107] indicated a nonchemical bonding of β-cyclodextrin and poly(vinylidene chloride). Due to the identity of the crystalline structure of monomeric adducts of β-cyclodextrin with both MMA and vinylidene chloride, on the one hand, and to the formation of an unstable product by a solid-state polymerization of β-cyclodextrin with MMA (dissociates into β-cyclodextrin and PMMA on being treated with hot water), on the other hand, these researchers also assumed an absence of chemical bonds in the stable product, obtained in the same way and consisting of poly(vinylidene chloride) and β-cyclodextrin.

The product of polymerization of β-cyclodextrin inclusion compounds, in contrast to the monomeric adduct, does not dissociate in either cold or hot water and is only partly soluble in DMF. With cyclodextrin and monomeric adducts washed off, the polymer contains 20 Wt.-% poly(vinylidene chloride) and has a molecular mass of 2×10^4. Calculations demonstrated that three moles of $-CH_2-CCl_2-$ links are bound with one mole of dextrin. The yield of such a product is high and amounts to 50%. The compound obtained does not dissociate into its components even after prolonged boiling in water and is capable of taking part in chemical reactions — β-cyclodextrin acetylation and poly(vinylidene chloride) dehydrochlorination, with the formation of new polyrotaxanes [100, 105].

Polyrotaxanes were studied by X-ray analysis [107]. Comparing the data on the specific volume of the polymeric and monomeric adducts of β-cyclodextrin poly(vinylidene chloride) and vinylidene chloride it was found that the values for polymers are the same or less than those for the low-molecular adducts. This is possible only when the polymer chain is within the cycle interior, as otherwise the specific volume would increase considerably.

Still another method for a directed synthesis of polyrotaxanes has been described in the literature [102, 109], namely a radical polymerization of various vinyl monomers under the action of azobutyronitrile in DMF solution in the presence of β-cyclodextrin. A detailed study of the behavior of these monomers with β-cyclodextrin in DMF demonstrated a "hot" crystallization to occur on heating of the solutions. Crystals represent compounds whose structure is identical to the structure of the same compounds specially prepared by mixing of individual solutions of the components. The fact that such inclusion compounds are formed on heating (the radical polymerization reaction proceeds at elevated temperatures) should favor the polymerization reaction. It turned out, however, that only homopolymers containing no β-cyclodextrin are formed in all cases, with the one exception of vinylidene chloride. Polymerization of vinylidene chloride in the presence of β-cyclodextrin yields, in addition to the homopolymer, also stable compounds containing 35 to 72 wt.-% β-cyclodextrin in polymers. This fact was attributed to the formation of crystalline inclusion compounds of β-cyclodextrin with various monomers in DMF solution at elevated temperatures [102]. As the temperature was further increased, the β-cyclodextrin content decreased gradually [109]. Polymerization of inclusion compounds of β-cyclodextrin with MMA and styrene, isolated beforehand, yielded polymeric adducts which contained considerable amounts of β-cyclodextrin and dissociated into their components on heating in water. Polymerization of the adduct of β-cyclodextrin and vinylidene chloride resulted in a product containing 85% cyclodextrin. The authors suggested [109] that the reaction of polymerization of monomeric adducts is in these cases preceded by the process of their association. This suggestion is supported by studies of the process of polymerization of the monomeric adduct of β-cyclodextrin and styrene in the presence of inclusion compounds of β-cyclodextrin with inactive molecules (such as benzene). If polymerization is preceded by association, then, when adducts of β-cyclodextrin with benzene are present in the system, they should also be included into such associates, thereby lowering the molecular mass of forming polymers. The molecular mass of polystyrene indeed decreases with increasing content of β-cyclodextrin-nonactive molecule inclusion compounds.

Thus, the studies of the directed synthesis of polyrotaxanes indicate the necessity of a preliminary ordering of the system (association, channel complexes of inclusion compounds). It should be pointed out that the above-described ways for the ordering are limited by a restricted choice of both macrocycles (β-cyclodextrin) and linear polymers (polyvinylidene chloride and its copolymer with allyl chloride [100, 107], and some polyamides [101]). There exists another possibility of ordering some macrocycles proposed and described by us [110]. We assumed that, similarly to linear oligoester acrylates [111, 112], some macrocycles are also capable of forming large ordered aggregates in complexing with a metal-containing compound. Polymerization of monomers in the presence of such aggregates should naturally result in products with a higher content of cycles on a chain compared to products of polymerization of the same monomers in the presence of individual cycles.

5 Synthesis of Polyrotaxanes with Elements of Directionality

5.1 Ordering of Macrocycles by Means of Complexing

For a successful synthesis of polyrotaxanes by polymerization of a monomer in the presence of macrocycles, ordered relative to one another, in the form of complexes of the latter with a metal-containing compound, the selected macrocycles should meet the following requirements: 1) large enough size; 2) molecules of the macrocycles should contain donor centres capable of coordination with a complexing agent — acceptor; and 3) the molecule should include rigid portions that would reduce the mobility of cycle fragments to such an extent so as to exclude a cooperative interaction of all or the majority of the donor atoms of the cycle with the ion of metal. In our opinion, the fulfilment of these requirements should result in that the metal ion arrangement to one side of the cycle interior, rather than in or over it, becomes preferable in the complexing. Binding two cycles into complexes by one ion also occurs, which makes possible the formation of swarm complexes consisting of many molecules. All this should promote the penetration of rings by the polymer chain in the process of polymerization.

The problems of synthesis of large cycles have been successfully solved and extensively described in the literature [113-116]. For a linear molecule penetration into the cycle interior without steric difficulties, the molecule should have the proper size. The data available in the literature are contradictory. For example, Harrison in his studies [2, 28] demonstrated that the yield of rotaxanes rises steadily up to a critical value (1.6%) with increasing size of the interior right up to a ring of 33 atoms, after which a decrease occurs due to the instability of the forming rotaxane; macrocyclanes as macrocycles and 1,13-di[tri-4-*tert*-butylphenylmethoxy]-tridecane as the "dumbbell" molecule were used. When studying the statistical synthesis of polyrotaxanes, Agam and Zilkha [99] found the degree of penetration of crown polyesters by polyethylene glycol PEG-600 to increase at a changeover from a 33- to a 58-atom cycle. Our studies [106] indicated, however, that when 34- and 40-atom cyclic urethanes are used as macrocycles, the degree of their threading on a growing polystyrene chain decreases with increasing size of the cycle interior. All these data clearly indicate a complexity of determining the criteria in selecting the required macrocycle size. We believe that a critical size of cycles exists and it depends on their chemical structure. When a critical size is exceeded, the macrocycle interior becomes inaccessible for penetration by a linear molecule because there are a large number of conformations with which a polymer chain cannot penetrate the cycles (such as "twisted" and "folded" conformers of cyclourethanes, shown in Fig. 3b and d). Moreover, the macrocycles with favourable conformation for penetration by a chain can easily "glide off" from macromolecules due to the large size of the cycle interior.

On the other hand, the chemical structure of the polymer chain penetrating into macrocycles is also essential for the formation of polyrotaxanes. If the size of a monomer unit exceeds the size of the macrocycle interior, the formation of polyrotaxanes becomes impossible, and if the monomer unit size is much less than that of the interior, the stability of polyrotaxanes is poor. A prerequisite for obtaining a stable polyrotaxane is the optimum ratio between the macrocycle interior size and the geometric

Fig. 3a–d. Molecular models of cyclourethanes: **a** — CU-A (oxygen atoms of ether groups **a** and alkyl oxygen atoms of acyloxy groups are in even conformations); **b** — "folded" conformer CU-B with similar arrangement of the same oxygen atoms; **c** — CU-A (arrangement of oxygen atoms of acyloxy and ether groups in odd conformations); **d** — twisted conformer — "figure-of-eight" CU-B

dimensions of the monomer unit — this ensures an adequate level of interaction with the "walls" of the macrocycle interior.

One of the prerequisites for the formation of complexes of cycles with a metal-containing compound is the presence in cyclic molecules of electron-donor atoms capable of forming coordination bonds with the metal ion. Despite there being a wide variety of heterocycles described in the literature, their complexing has been investigated mostly for cyclic polyesters (crown polyesters) [37,113-123]. These attracted attention because of their unusual behavior at coordination bonding, which consists of the manifestation of a cooperative interaction of all, or the majority, of donor atoms with a metal ion, the conformation of a molecule in the complex being such that its donor atoms embrace the metal ion like a crown and hold it strongly in or above the interior. The complexing properties of crown-shaped macrocycles is based on the correspondence between the size and shape of the macrocycle interior and the size of radii of metal ions, owing to which a "molecular recognition", showing up in a selective formation of inclusion compounds with the ions, is ensured [18,19,124]. The stability of complexes with metal ions depends on many factors: the nature of the cation, of the corresponding anion, and of the solvent [125]. Their stability is so high that displacing the cation out of the interior by a neutral molecule is extremely difficult.

Macrocycles of such a type have been studied in detail and are extensively employed in various fields of chemistry [119,124-131].

The specific features of complexing between crown polyesters and metal cations indicate the impossibility of using such complexes for a directed synthesis of polyrotaxanes. Only their statistical synthesis with the use of free macrocyclic polyethers is feasible [99].

Obtaining ordered aggregates of macrocycles by complexing with metal salts requires that the effect of a cooperative binding of donor atoms in polydentant cyclic molecules, in contrast to crown-shaped cycles, can not take place. This becomes possible when cyclic molecules contain fragments preventing them from taking the conformation of crowns. Introduction of some groups (biphenyl [132]) binaphtyl [123], urethane [133,134]) reduces the flexibility of a macrocycle molecule, which makes increases the participation of donor atoms independently of one another in the complexing. If, moreover, the complexing metal is capable of coordinating at least two donor atoms, swarm complexes can arise.

The requirements placed upon macrocycles with the aim of their organization into ordered swarms and their further use for obtaining polyrotaxanes are [103], met by urethane macrocycles based on hexamethylene diisocyanate and di- and triethylene glycols (CU-A and CU-B respectively). They are large enough (consisting of 34 and 40 atoms) for penetration of a linear molecule into their interior, have four rigid urethane groups and several types of donor atoms (oxygen atoms of ether, alkoxy-, and carbonyl groups, nitrogen atoms of urethane groups) [133,134]:

Zinc chloride was used as the complexing agent which forms tetrahedral complexes [110].

The complexing was studied by ^{1}H-NMR and viscosimetry [110,136] and using the cyclourethane modeling results [137].

The NMR technique indicated that adding $ZnCl_2$ to cyclourethanes gives rise to a tendency for an increase of the chemical shift of all signals. The maximum changes of the chemical shifts in the complexing were observed for signals of protons of methylene groups which are adjacent to oxygen atoms of ether groups $-CH_2-O-CH_2-$ and group $-CH_2-O-C(O)-$. The obtained results indicate a $ZnCl_2$ coordination predominantly with oxygen atoms of cyclic urethanes. The absence of zinc ion interaction with nitrogen atoms of an urethane group is likely to stem both from a lower electron-donor ability of a nitrogen atom compared to that of an oxygen atom [135] and from binding of NH groups of cyclourethanes by dimethyl sulfoxide in which the complexing was studied [110].

The molar ratio method demonstrated the formation of complexes of a stoichiometric composition 1:1. With this stoichiometry, an existence of complexes where the C_{CU}/C_{ZnCl_2} ratio is of 1:1; 2:2, 3:3, etc. may be assumed [110, 136]. The increases in the viscosity of solutions of the complexes over that of solutions of the initial cyclourethanes also demonstrate the formation of large aggregates.

Modeling of the urethane macrocycles was conducted in order to elucidate the structure of the complexes [137]. Constructing the Steward-Brigleb models of molecules of both cycles made it possible to demonstrate visually that a wide set of various conformers is characteristic for the urethane cycles. Depending on the relative arrangement of atoms in the molecule, the cycle configuration can vary from a more stretched (Fig. 3a) to a less stretched (Fig. 3c) ellipse. "Folded" conformers (Fig. 3b) can be formed as a result of a possible formation of intramolecular hydrogen bonds at an appropriate arrangement of carbonyl and NH-groups. An increase in the cycle size at a change from CU-A to CU-B results in a higher mobility of cycle fragments, which can show up in the formation of conformations twisted in the form given in Fig. 3d. Assuming the existence of cyclourethanes as a mixture of various conformers and a predominantly tetrahedral shape of Zn^{2+} complexes [138-140], an attempt was undertaken to represent the structure of the complexes formed. It was demonstrated [137] that the binding of a zinc ion, in a complex with oxygen atoms, separately in every fragment of cyclourethanes with the conformation shown in Fig. 3a would be difficult due to a low probability of a cooperative interaction of the donor atoms. A cooperative interaction of all the oxygen atoms of the conformer in question is impossible due to an insufficient length of methylene units, which does not allow the atoms to come close enough to interact with the zinc ion. The oxygen atoms of the other conformers are directed to different sides of the cycle, and therefore their cooperative binding is altogether impossible (Fig. 3c).

The origination of complexes through the formation of coordination bonds between Zn^{2+} and two oxygen atoms of any type in a cyclourethane molecule is also unlikely, since either the atoms are too far from each other distortions of the tetrahedral angle

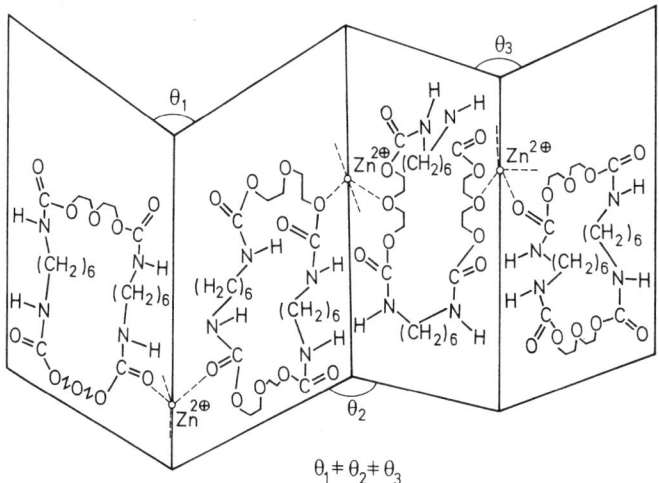

Fig. 4. Schematic of "swarm" complex of CU-A with $ZnCl_2$

O ... Zn ... O are needed for the appearance of such bonds. It is reasonable to assume that the formation of donor-acceptor bonds of Zn^{2+} occurs with oxygen atoms of different cyclic urethane molecules.

Thus, it can be concluded that the 1:1 stoichiometric composition can correspond to complexes $[(ZnCl_2)_n(CU)_n]$. The complexes can be represented as aggregates of CU molecules interconnected by zinc ions, where zinc can coordinate oxygen atoms of any type. Angles characteristic of a tetrahedral coordination of ligands around the central atom form between the Zn ... O bonds. Such a complex is schematically shown in Fig. 4 [141].

5.2 Synthesis of Polyrotaxanes with the Use of Swarm Complexes and Investigation of Their Structure

To determine the efficiency of ordering of macrocycles in a swarm complex and its effect on polyrotaxane synthesis, we studied styrene polymerization in the presence both of a complex of cyclourethanes with $ZnCl_2$ (method with elements of directionality) and of cyclic urethanes (statistical method) [142]. The structure of the compounds obtained has been examined. A specific feature of these syntheses was an absence of specially added agents for the initiation of radical polymerization. The reaction was initiated by traces of air oxygen present in the system. Such an initiation method resulted in a very slow rate and promoted formation of macromolecules of a high molecular mass (10^5 to 10^6). Polymerization proceeded under heterogenous conditions since CU and their complexes are insoluble in styrene but swell in it. The equilibrium degree of swelling of CU-A in styrene is of 150%, and of CU-B 50% [106].

Conducting the polymerization under such conditions allowed us to assume the following mechanism of polyrotaxane formation. First the swelling of cyclic urethanes in the monomer occurs, i.e. styrene penetrates into the bulk of cyclourethanes or their complexes also including their interiors. Polymerization of styrene which is both in the interior and on the outside results in the formation of a polymer whose molecules "pierce" the cycles, i.e. the possibility of formation of compounds with a polyrotaxane structure arises.

Since the polymerization resulted in a polymer with a high molecular mass, it was assumed that the length of polymer chains "piercing" the macrocycles prevents the "gliding-off" of rings from them, and therefore no comonomers with large bulky substituents were used. Gliding-off of rings is probable only at the chain ends in such compounds. Desired products were isolated from such systems by extraction of the complexing agent, free homopolymers and macrocycles. IR spectra of the isolated compounds consist of absorption bands characteristic for polystyrene (PS) and CU. Studying the X-ray scattering of the same compounds at wide angles demonstrated that the scattering functions by samples synthesized by styrene polymerization in the presence of both cyclourethane — $ZnCl_2$ complexes and unbound cycles have the form of curves with maxima characteristic for scattering by a free PS and unbound urethane cycles (Fig. 5).

To find out whether or not CU is chemically bound with a growing PS chain, styrene polymerization in the presence of a linear analog, diurethane obtained from hexamethylene diisocyanate and ethylene glycol monoethyl ether, was studied [143, 144].

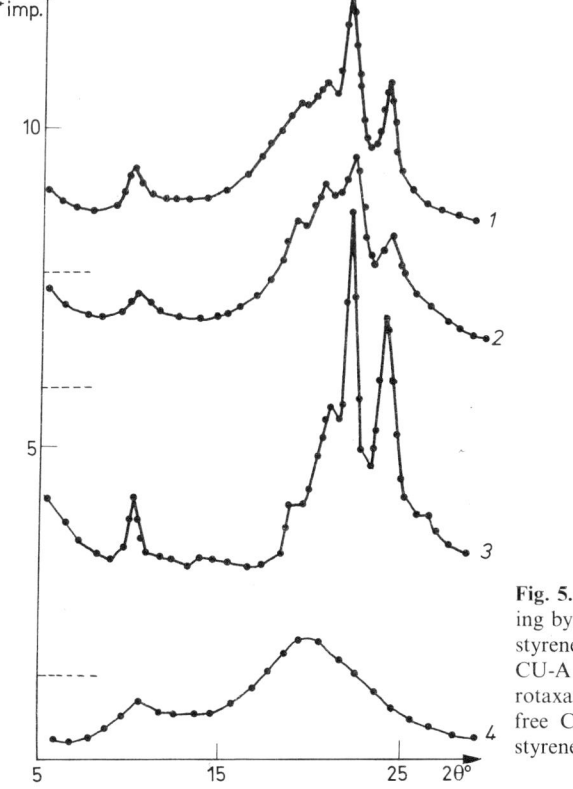

Fig. 5. Diffractogram of X-ray scattering by: *1* — polyrotaxane obtained by styrene polymerization in presence of CU-A — $ZnCl_2$ complex; *2* — polyrotaxane synthesized from styrene and free CU-A; *3* — CU-A; *4* — polystyrene

The X-ray diffraction of the product after PS extraction turned out to be similar to the X-ray scattering by the initial diurethane. These results prove an absence of a chemical interaction between urethane and PC. The identical nature of CU and their linear analog made it reasonable to suggest an absence of the chemical bonding also in the case of PS and CU. It follows that a mechanical engagement between cyclourethanes and PS molecule occurs in the samples, i.e. polyrotaxane-structure compounds are formed.

Diffractograms of polyrotaxanes (Fig. 5) indicate a nonuniform arrangement of molecules along the linear polymer chain. Since a definite degree of cyclourethane crystallinity is retained in the structure of the forming polyrotaxane, it should be assumed that the molecule has portions with a higher content of cycles and portions where cycles are absent. This may be schematically represented as follows:

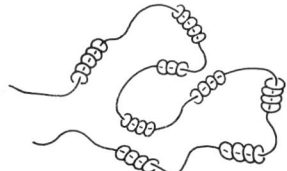

Such a local arrangement of cycles along the chain is typical both for a statistical polyrotaxane and for a polyrotaxane prepared by styrene polymerization in the presence of swarm complexes of cyclourethanes with $ZnCl_2$. For the latter, portions with a higher concentration of cycles can be attributed to the swarm structure of complexes, while for a statistical polyrotaxane such a structure appears to be due to that the initial cyclourethanes are oriented with respect to one another into ordered formations with the aid of intermolecular hydrogen bonds. The length of the formations is shorter than in swarm complexes.

The composition of polyrotaxanes was estimated from the data of the X-ray scattering by the polyrotaxanes, initial cyclourethanes, and PS [106]; the results are given in the Table 1.

Table 1. Compositions of CU and PS-based polyrotaxanes

Polyrotaxanes[a]	Volume fraction of CU in polyrotaxanes, %	Number of monomeric units of PS per one cycle
PR-A1	40	7 ± 1
PR-A2	20	19 ± 1
PR-B1	35	12 ± 1
PR-B2	25	20 ± 1

[a] PR-A1, PR-B1 — polyrotaxanes produced by styrene polymerization in the presence of complexes of $ZnCl_2$ with CU-A and CU-B respectively; PR-A2, PR-B2 — polyrotaxanes synthesized in the presence of initial CU-A and CU-B respectively

From the data in Table 1 it follows that styrene polymerization in the presence of cyclourethane — $ZnCl_2$ complexes results in products (PR-A1 and PR-B1) where the degree of penetration of rings by the growing chain is higher than in polyrotaxanes produced by the statistical method (PR-A2 and PR-B2). Such a difference can be explained by the complexes of $ZnCl_2$ with cyclic urethanes being more lengthy ordered systems than the initial urethane macrocycles.

Comparing the results of synthesis of polyrotaxanes based on the two cyclourethanes demonstrated that the degree of penetration of rings by the growing polymer chain in the polyrotaxanes tends to decrease by passing from CU-A to CU-B. For polyrotaxanes produced by the statistical method (PR-A2 and PR-B2) no changes are practically observed, but for the polymers synthesized in the presence of complexes with $ZnCl_2$ the difference is marked (PR-A1 and PR-B1) [106].

The results show good agreement with the data on complexing and modeling of the initial cycles and their complexes. The study and preparation of models of the synthesized polyrotaxanes correspond to the presented data [137].

Studying the molecular models of polyrotaxanes for various CU conformers indicated that a large interior of CU-B favors the arrangement of the styrene molecule benzene ring plane in the cycle plane (Fig. 6). Such an arrangement of the monomer results in reduced steric hindrances for polymerization, i.e., it is a favorable factor in the synthesis of polyrotaxanes. At the same time, however, the distance between

 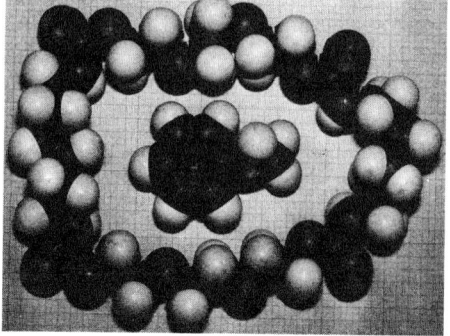

Fig. 6a and b. Molecular models of polyrotaxanes based on polystyrene and **a** CU-A, **b** CU-B

a styrene molecule and the "inside walls" of the cycle interior is greater, with the result that the interaction due to Van der Waals forces decreases. The monomer molecule is more weakly bound in the CU-B interior, which will increase the probability of "gliding-off" of rings in styrene polymerization. In addition, due to a possible CU-B existence in the "figure-of-eight" conformation as well as to a greater number, as against CU-A, of conformers "folded" because of intramolecular hydrogen bonds, the probability of formation of polyrotaxane structures in a system with CU-B is lower. Such assumptions are supported by studies of swelling of macrocycles in styrene (the degree of swelling of CU-A in styrene is higher by a factor of 3 compared to CU-B) [105]. One of the causes of such behavior of cyclourethanes — swelling in a monomer — is a lower hydrophobity of CU-B and arrangement of some CU-B molecules in such conformations which the monomer molecules cannot penetrate into the cycle interior. It is the action of these mutually exclusive factors that accounts for no changes in the degree of penetration of rings by the chain observed at a change of the cycle site.

A higher degree of penetration for polyrotaxanes produced in the presence of the CU-A — $ZnCl_2$ complex compared to CU-B — $ZnCl_2$ complexes [106] is closely linked to the complexing of cyclourethanes with $ZnCl_2$. A greater number of cyclic molecules in the complex and a greater number of CU-A complexes result in a greater degree of penetration of these cycles by a polystyrene molecule. Moreover, the process of threading the rings on the chain is [145], affected by the solvent. Dimethyl sulfoxide (DMSO) breaks the H-bonds in the initial urethane, including also the intramolecular ones. This may result in a partial opening of the interior in folded conformations of the cycle. It can also be supposed that zinc "fixes" the structure of cycles in the forming CU — $ZnCl_2$ complexes, so that the structure can also remain after the removal of the solvent; this also increases the number of penetrated cycles.

Thus, the degree of penetration for polyrotaxanes produced in the presence of "swarm" complexes depends on the size of aggregates of complexes and their number in the system as well as on the possibility of "opening" the cycle interior under the action of DMSO. The degree of threading of rings on growing polymer chains rises with the increasing number of CU molecules in a swarm and number (concentration) of swarm complexes.

6 Conclusion

The synthesis and study of polycatenanes and polyrotaxanes is of interest from various standpoints. First, the search for new topological compounds, exhibiting non-trivial properties, is essential. The existence of such substances in nature and their synthesis in laboratories made it possible to discover a new construction principle. Structural features of the compounds under consideration may give rise to their unusual properties. Second, polycatenanes and polyrotaxanes are a new class of polymers. Mixing dissimilar polymers for creating novel materials offering a valuable set of properties has become one of the leading trends in polymer materials science. It has been, however, well established that mixing of polymers is not accomplished at the molecular level, but mixtures represent a microheterogenous colloidal system [42]. Polycatenanes and polyrotaxanes allow the "mixing" of polymer chains at the molecular level and thereby allow us to change considerably the structure and properties of a material. Topological entanglements with a definite lifetime are known [146] to exist in melts of polymers and of their mixtures. At strain, the entanglements play the part of junctions in a spatial network. In ordinary polymer systems, however, a creeping of chains linked by topological entanglements may occur, whereas in the case of polyrotaxanes and polycatenanes quite different behavior of such molecules and a different set of elastic properties may be expected, which give rise to a number of new physical problems.

Examination of the literature indicates that several methods of polyrotaxane synthesis have been studied. Only one method for preparation, the statistical one, is so far known for polycatenanes, while three such methods — statistical, directed, and a method with elements of directionality — exists at present for polyrotaxanes and the advantage of directed methods, whose prerequisite is an ordering of the initial mixture, has been found. It is obvious that the described ways for ordering of systems where macromolecular topological compounds can form, are not yet exhausted. A search for new techniques of their ordering will open new possibilities in the synthesis of various topological polymers.

It is difficult to forecast the ways for a practical application of the new types of polymeric compounds — polycatenanes and polyrotaxanes — which to a certain extent is due to the difficulties of their synthesis. It can be, however, expected that a detailed study of their physical and mechanical properties, which will open the ways for their practical application, will become possible as the synthetic techniques are developed and simpler ways for the synthesis are found.

The present review has attempted to cover comprehensively those issues of chemistry and physical chemistry of topological macromolecular compounds which have been presented up to now in the literature world wide, and the authors express the hope that fundamental studies in this field will be successfully continued.

7 References

1. Frisch HL, Wassermann E (1961) J. Amer. Chem. Soc. 83: 3789
2. Harrison JT (1974) J. Chem. Soc., Perkin Trans., pt. 1 2: 301
3. Ben-Efraim DA, Batich C, Wasserman E (1970) J. Amer. Chem. Soc. 92: 2133
4. Wolowsky R (1970) J. Amer. Chem. Soc. 92: 2132

5. Schill G, Zollenkopf H (1969) Lieb. Ann. Chem. 721: 53
6. Schill G (1971) Catenanes, rotaxanes, knots, Academic press, New York
7. Schill G, Doerjer G, Logemann E, Fritz H (1979) Chem. Ber. 112: 3603
8. Schill G, Zürcher C (1977) Chem. Ber. 110: 2046
9. Logemann E, Schill G, Vetter W (1978) Chem. Ber. 111: 2615
10. Schill G, Doerjer G, Vetter W (1980) Chem. Ber. 113: 3697
11. Schill G, Ortlieb H (1981) Chem. Ber. 114: 877
12. Schill G, Zürcher C (1980) Chem. Ber. 113: 2052
13. Schill G, Schweickert N, Fritz H, Vetter W (1983) Angewan. Chem. 95: 909
14. Schill G, Logemann E, Litke W (1984) Chem. unserer Zeit 18: 130
15. Nakedzaki M, Yamamoto K (1984) Chemistry, v 39, Kagaku, p 517
16. Rissler K, Schill G, Fritz H, Vetter W (1986) Chem. Ber. 119: 1374
17. Agam G, Zilkha A (1976) J. Amer. Chem. Soc. 98: 2514
18. Barabanov VA, Davydova SA (1982) Vysokomolekul. soed., ser. A, 24: 899
19. Mathias LJ, Al-Jumah KB (1979) Polymer News 6: 9
20. Callahan D, Frisch HL, Klempner D (1975) Polym. Eng. and Sci. 13: 70
21. Maciejewski M (1980) Prace nauk. Warsz. Chem. N 20: 3
22. Jacobson H (1984) Macromolecules 17: 705
23. Rigbi Z, Mark JE (1986) J. Polym. Sci., pt. B 24: 443
24. Sokolov VI (1973) Uspekhi khim 42: 1037
25. Kostjanovskij RG (1965) Khimija i zhizn' 2: 35
26. Sokolov VI (1982) Vvedenie v teoreticheskuju stereokhimiju, Nauka, Moscow, p 174
27. Dmitriev IS (1980) Molekuly bez khimicheskikh svjazej, Khimija, Moscow, p 95
28. Harrison JT (1972) J. Chem. Soc. Chem. Commun. 4: 231
29. Shul'pin GB (1979) Khimija i zhizn' 4: 21
30. Slenk W (1949) Lieb. Ann. Chem. B. 565, S. 204
31. Cramer F (1954) Einschluss-Verbindungen, Springer, Berlin Göttingen Heidelberg
32. Lehn JM (1978) In: Itogi i perspektivy razvitija bioorganicheskoj khimii i makromolekuljarnoj biologii, Nauka, Moscow, p 41
33. Albano VG, Sansoni H, Chini P, Martinengo S, Strumolo D (1976) J. Chem. Soc., Dalton Trans 1: 970
34. Johnson BFG, Johnston RD, Lewis J (1968) J. Chem. Soc. A: 2865
35. Novikov JN, Vol'pin ME (1971) Uspekhi khim 40: 1568
36. Mandelcorn L (ed) (1964) Non-stoichiometric compounds, Academic, New York
37. Pedersen KD, Frensdorff CK (1972) Angewan. Chem., Int. Ed. 11: 16
38. Brown JEJ, White DM (1960) J. Amer. Chem. Soc. 82: 5671
39. White DM (1960) J. Amer. Chem. Soc. 82: 5678
40. Lehn JM, Sauvage JP (1975) J. Amer. Chem. Soc. 97: 6700
41. Maciejewski M (1982) J. Macromol. Sci. A17: 689
42. Lipatov YS (1984) Colloid Chemistry of Polymers. Elsevire (1988)
43. Wasserman E (1960) J. Amer. Chem. Soc. 82: 4433
44. Schill G, Lüttringhaus A (1964) Angewan. Chem. 76: 567
45. Schill G, Herschel R (1970) Lieb. Ann. Chem. 731: 113
46. Schill G, Zürcher C, Vetter W (1973) Chem. Ber. 106: 228
47. Schill G, Rißler K, Fritz H (1983) Chem. Ber. 116: 1866
48. Schill G, Beckmann W, Vetter W (1980) Chem. Ber. 113: 941
49. Schill G, Tafelmair L (1971) Synthesis 10: 546
50. Porzi G, Concilio C, Bongini A (1973) Gazz. Chim. Ital. 103: 393
51. Harrison JT, Harison S (1967) J. Amer. Chem. Soc. 89: 5723
52. Dietrich-Bucheker CO, Sauvage JP, Kintzinger JT (1983) Tetrahedron Lett. 24: 5095
53. Ogino H (1981) J. Amer. Chem. Soc. 103: 1303
54. Yamari K, Shimura I (1983) Bull. Chem. Soc. Jap. 56: 2283
55. Dietrich-Bucheker CO, Sauvage JP, Kern JM (1984) J. Amer. Chem. Soc. 106: 3043
56. Cesario H, Dietrich-Bucheker CO, Guilhem J, Bescal C, Sauvage JP (1985) J. Chem. Soc. Chem. Commun. 5: 244
57. Guilhem J, Cesario M, Dietrich-Bucheker CO, Sauvage JP, Kintzinger JP, Pascard C, Rich C (1985) 9th European Crystallographic Meeting, Torino, part 2, p 127

58. Sauvage JP (1985) Nouveau Jornal Chim. 9: 299
59. Dietrich-Bucheker CO, Sauvage JP, Weiss TJ (1986) Tetrahedron Lett. 27: 2257
60. Sauvage JP, Weiss J (1985) J. Amer. Chem. Soc. 107: 6108
61. Ogino H, Ohata R (1984) Inorgan. Chem. 23: 3312
62. Yamanari K, Shimura J (1984) Bull. Chem. Soc. Jap. 57: 1596
63. Freudenberg K, Meyer-Deliys M (1938) Chem. Ber. 71: 1596
64. Maciejewski M; Patent PRL No. 100,828, 11.05.74—30.04.79
65. Maciejewski M; Patent PRL No. 103,276, 29.12.75—31.10.79
66. Cramer F, Henglein FM (1957) Chem. Ber. 90: 2561
67. Cramer F, Saenger W, Statz H-C (1967) J. Amer. Chem. Soc. 89: 14
68. Hybl A, Rundle PE, Williams DE (1965) J. Amer. Chem. Soc. 87: 2779
69. Demarco RV, Thakkar AL (1970) J. Chem. Soc. Chem. Commun. 1: 2
70. Lüttringhaus A, Cramer F, Prinzbach H (1957) Angew. Chem. 69: 137
71. Lüttringhaus A, Cramer F, Prinzbach H, Henglein FM (1958) Lieb. Ann. Chem. 613: 185
72. Dubous J-E, Ranaye A (1975) Bull. Soc. Chim. France, part 2, 5–6: 1401
73. Hudson B, Vinograd J (1967) Nature 216: 647
74. Glauton DA, Vinograd J (1967) Nature 216: 652
75. Hudson B, Vinograd J (1969) Nature 221: 332
76. Wang JC (1970) Biopolymers 9: 489
77. Chem. Eng. News (1970) 48: 46
78. Simpson L, Da Silva A (1971) J. Mol. Biol. 56: 443
79. Agol VI, Romanova LI, Tol'skaja EA, Bogdanov AA (1969) Dokl. Akad. Nauk SSSR, 189: 428
80. White JH, Cozzarel NR (1984) Biological Sci 81: 3322
81. Frank-Kamenetskij MD (1981) Khimija i zhizn' 9: 27
82. Spengler SJ, Stasiak A, Cozzared NR (1985) Cell 42: 325
83. Schill G, Zürcher C (1971) Naturwissenschaften 58: 40
84. Wasserman E (1962) Scientific American 207: 94
85. Burdick GD (1970) Naturwissenschaften 57: 245
86. Klempner D, Frisch HL (1970) J. Polymer Sci., pt. B, 8: 525
87. Klempner D, Frisch HL, Frisch KC (1971) J. Elastoplast. 3: 2
88. Klempner D, Frisch HL, Frisch KC (1970) J. Polym. Sci., pt. A2, 8: 921
89. Frisch HL, Klempner D, Frisch KC (1969) J. Polym. Sci., pt. B, 7: 775
90. Frisch HL, Klempner D, Frisch KC, Kwel TK (1970) Amer. Chem. Soc., Polym. Prepr. 11: 483
91. Lipatov YS, Sergeeva LM (1979) Vzaimopronikajushchie polimernye setki, Naukova Dumka, Kiev, p 44
92. Karagounis G, Pandi-Agathokli J (1970) Pract. Acad. Athenon, 45: 118
93. Karagounis G, Pandi-Agathokli J (1972) Pract. Panelleiou Chem. Synedriou, 4th 2: 213
94. Karagounis G, Kontaraki E, Petassis E (1973) Pract. Acad. Athenon 48: 197
95. Karagounis G, Pandi-Agathokli J, Petassis E, Alexakis A (1973) Folia Biochim. Biol. Graeca 10: 31
96. Karagounis G, Pandi-Agathokli J, Kontaraki E (1972) Chem. Chrom. 1: 103
97. Karagounis G, Pandi-Agathokli J, Konthraki E (1976) Proc. Int. Conf. Colloid and Surface Sci. vol 1, Budapest, Acad. Kiodo, 1975, p 671
98. Karagounis G, Pandezi M (1976) Proc. 5th Int. Conf. Raman Spectrosc., Freiburg, Freiburg im Breisgau, 1976, p. 72.
99. Agam G, Graiver D, Zilkha A (1976) J. Amer. Chem. Soc. 98: 5206
100. Maciejewski M, Smets G (1975) Prace nauk. Inst. Technol. organicz. i tworzyw cztuczn. 16: 57
101. Ogata N, Sanui K, Wada J (1976) Polym. Lett. Edit. 14: 459
102. Maciejewski M, Gwizdowski A, Peczak P, Pietrzak A (1979) J. Macromol. Sci A 13: 87
103. Lipatova TE, Kosyanchuk LF, Shilov VV (1985) J. Macromol. Sci. A22: 361
104. Maciejewski M (1979) J. Macromol. Sci. A 13: 1175
105. Maciejewski M, Durski Z (1981) J. Macromol. Sci. A 16: 441
106. Lipatova TE, Kosyanchuk LF, Shilov VV, Gomza YP (1985) Vysokomolekul. soed. A 27: 556
107. Maciejewski M (1979) J. Macromol. Sci. A 13: 77
108. Maciejewski M, Panaciewiec M, Jarminska D (1978) J. Macromol. Sci. 12: 701

109. Maciejewski M (1979) Iupac Macro Mainz: 26th Int. Symp. Macromol., Mainz prepr. Short. Commun. 1: 60
110. Lipatova TE, Kosyanchuk LF, Khramova TS (1982) Dokl. Akad. Nauk Ukr. SSR B 3: 35
111. Lipatova TE, Budnikova VA, Siderko VA (1965) Vysokomolekul. soed. 7: 580
112. Lipatova TE (1974) Kataliticheskaja polimerizatsija oligomerov i formirovanie polimernykh setok, Naukova Dumka, Kiev, p 90
113. Hiraoka M (1982) Crown compounds, Kodansha, Tokyo
114. Bogatskij AV, Lukjanenko NG, Kirichenko TI (1985) Zhurn. VKhO im. D. I. Mendeleeva 30: 487
115. Markovskij LN, Kal'chenko VI (1985) Zhurn. VKhO im. D. I. Mendeleeva 30: 528
116. Voronov MG, Knutov VI (1985) Zhurn. VKhO im. D. I. Mendeleeva 30: 535
117. Izatt M, Christensen JJ (1981) Progress in macrocyclic chemistry, parts 1, 2 Wiley, New York
118. Pedersen CJ (1967) J. Amer. Chem. Soc. 89: 7017
119. Bogatskij AV, Lukjanenko PG (1981) Sinteticheskie makrotsikly — novye reagenty nastojashchego i budushchego, Znanie, Kiev, p 48
120. Black DS, Hartshort AJ (1973) Coord. Chem. Rev. 9: 219
121. Farago HE (1977) Inorg. Chim. Acta 25: 71
122. Tusek-Bozić LJ, Danesi PR (1979) Inorg. and Nucl. Chem. 41: 833
123. Kram DJ, Heegenson RS, Sausa LR, Timko DM, Newcom M, Moro P, de Yong F, Gokel GV, Hoffman DH, Domeyer LA, Madan K, Kaplan L (1975) Khim. geterotsiclicheskikh soed. 10: 1299
124. Davydova SL, Barabanov VA (1980) Koordinats. khim 6: 823
125. Frensdorff HK (1971) J. Amer. Chem. Soc. 93: 600
126. Lukjanenko NG, Bogatskij AV, Kirichenko TI (1985) Zhurn. VKhO im. D. I. Mendeleeva 30: 571
127. Bogatskij AV, Nazarov EI, Golovenko NJ (1985) Zhurn. VKhO im. D. I. Mendeleeva 30: 593
128. Laskorin BN, Jakshin VV (1985) Zhurn. VKhO im. D. I. Mendeleeva 30: 579
129. Zolotov JA (1985) Zhurn. VKhO im. D. I. Mendeleeva 30: 584
130. Makumovich AG, Verizhnikov AV, Safitov VN (1978) Khimija i tekhnologija elementoorg. soed. i polimerov (Kazan') 7: 37
131. Kainuma H, Naito K, Hirai H (1979) Amer. Chem. Soc. Polym. Prepr. 20: 970
132. Nishikido J, Inasu T, Yoshiro T (1973) Bull. Chem. Soc. Jap. 46: 263
133. Heitz W, Höcher H, Kern W, Ullner H (1971) Macromol. Chem. 150: 73
134. Kosyanchuk LF, Lipatova TE (1982) XIV Ukr. respubl. konf. po org. khimii — Tezisy dokl., Odessa p 130
135. Bakalo LA, Chirkova LI, Lipatova TE (1982) Zh. org. khim., 18: 1416
136. Lipatova TE, Kosyanchuk LF, Khramova TS (1983) Teor. eksperim. khim. 19: 323
137. Lipatova TE, Kosyanchuk LF (1986) Teor. eksperim. khim., 22: 507
138. Orgel L (1961) An Introduction to Transition-Metal Chemistry, Wiley, New York
139. Noltes JG, Boersma J (1967) J. Organometal. Chem. 7: 6
140. Slabauygh WH, Parsons T (1976) General Chemistry, Wiley, New York
141. Lipatova TE, Kosyanchuk LF (1985) XV Vsesojuzn. Chugaevskoe soveshchanie po kompleksnym soed. — Tez. dokl., pt. 1, KGU, Kiev, p 74
142. Lipatova TE, Kosyanchuk LF (1985) XXII konf. po vysokomolekul. soed., Alma-Ata, Tez. dokl., p 98
143. Lipatova TE, Kosynchuk LF, Gomza YP, Shilov VV, Lipatov YS (1982) Dokl. Akad, Nauk SSSR 262: 1379
144. Lipatova TE, Kosyanchuk LF (1984) Pjataja respubl. konf. po vysokomolekul. soed. — Tez. dokl., Naukova Dumka, Donetsk, p 18
145. Kosyanchuk LF (1984) Avtoref. dis. na soiskanie uch. st. kand. khim. nauk, IKhVS AN Ukr. SSR, Kiev
146. de Gennes P (1979) Scaling concepts in polymer physics, Cornell University Press, Ithaca London

Editor: K. Dušek
Received May 10, 1988

Reactions and Photodynamics in Polymer Solids

Kazuyuki Horie and Itaru Mita
Research Center for Advanced Science and Technology, The University of Tokyo,
4-6-1 Komaba, Meguro-ku, Tokyo 153, Japan

Reactions in amorphous polymer solids are first characterized in terms of influences of molecular motion of matrix polymers and non-homogeneity of reaction sites. Specific features of photophysical processes, photoisomerization, photodimerization, chain scission and crosslinking reactions in polymer solids are then discussed separately.

1 Introduction 79

2 Historical Background 79

3 Characteristics of Reactions in Amorphous Polymer Solids 82
  3.1 Factors Affecting Reactions in Polymer Solids 82
  3.2 Effects of Molecular Motion and Temperature 83
  3.3 Heterogeneous Progress of Reactions 86
      3.3.1 Deviation from First-Order Kinetics 86
      3.3.2 Photoreactions and Photophysical Processes 87
      3.3.3 Effects of Binding to the Polymer Chain on the Reactivity of
           Chromophore 88
  3.4 Kinetic Analysis of Non-Homogeneous Reactions 89

4 Photophysical Processes 90
  4.1 Non-Exponential Phosphorescence Decay in Polymer Solids 91
  4.2 Phosphorescence Decay of Benzophenone in Polymer Solids 93
  4.3 Kinetics for Non-Exponential Decay Due to Dynamic Quenching . . . 95
  4.4 Hydrogen Abstraction of Benzophenone Triplet in Poly(vinyl alcohol) . 98
  4.5 Phosphorescence Decay of Benzophenone under Multi-Photon Conditions 99

5 Isomerization Reactions 101
  5.1 Spiropyran and Other Photochromic Compounds 101
  5.2 Azobenzene and its Derivatives 106
  5.3 Photochemical Hole Burning 111

6 Photodimerization 115

7 Chain Scission and Crosslinking Reactions 118
   7.1 Norrish Type I and Type II Reactions 118
   7.2 Degradation of Vinyl and Aromatic Polymers 120

8 Acknowledgement 125

9 References 125

1 Introduction

Reactions in polymer solids greatly differ from those in solutions mainly because the mobility of reactants is far more suppressed in the former than in the latter [1-4]. Another important difference is that reactions [3-5] in polymer solids frequently proceed heterogeneously owing to the microscopically heterogeneous state of aggregation or the free volume distribution. Polymer matrices also cause physical quenching and the chemical reaction of doped photo-excited chromophores. Thus, even in an organic glass, poly(methyl methacrylate) (PMMA), which is usually considered an inert matrix for photophysical and photochemical processes, the decay of benzophenone phosphorescence is markedly different from a simple exponential [3,4].

Typical chemical reactions in polymer solids are degradation and crosslinking induced by heating, photo or γ-ray irradiation as well as photodimerization and photoisomerization, the latters being used as the key reactions in photoimaging systems. Industrial developments in photography, lithography, and photocopying depend to a great extent on photochemistry in macromolecular media. Early studies of photochemistry in polymer solids chiefly aimed at the macroscopic physical performance of photosensitive polymeric materials, but efforts [1-5] were also made toward understanding the basic features of photoreactions in such systems at a microscopic or molecular level. Transient measurements of the reactions and photodynamics in polymer solids have become popular for the understanding of real-time reaction rates and reaction intermediates with the advent of laser flash and pulse radiolysis techniques [6,7]. Topochemical reactions in the crystalline state such as photocycloaddition of diolefines, diacetylenes, and other unsaturated compounds [8-10] are also important solid-state reactions, but, in what follows, we will be concerned mainly with the reactions in amorphous polymers.

In this review article, after a description of the historical background of the present subject, some general features of reactions in amorphous polymer solids are summarized and then specific features of photo-physical processes, photoisomerization, photodimerization, chain scission, and crosslinking reactions are discussed separately. Typical theoretical treatments of solid-state reactions are reviewed in relation to the interpretation of experimental results.

2 Historical Background

Mechanical and other physical properties of amorphous polymers show marked changes in the glass-rubber transition region. The glass transition temperature T_g is generally defined as a temperature at which large-scale rotational and translational motion (microbrownian motion) of the polymer backbone segments begin to be frozen-in. Thus it was suggested that no reaction occurs in amorphous polymers below T_g. The radiation-induced solid-state polymerization and the cessation of polymerization due to vitrification may serve as typical examples of this suggestion.

The polymerization of trioxane in the crystalline state is a well-known example of radiation-induced solid-state polymerization [11]. In the polymerization of crystalline monomers, the topotactical conditions are very important and polymerization is possible only for a very few monomers. The situation is completely different for amorphous solid monomers. It was sometimes claimed [12] that polymerization actually proceeded in the glassy state of a monomer during irradiation. However, Nakatsuka et al. [13,14] showed by low temperature differential thermal analysis (DTA) of the irradiated vinyl acetate monomer in the glassy state, that polymerization took place, not during the irradiation, but during the subsequent heating at a temperature just above T_g (see the exothermic peaks for irradiated samples in Fig. 1. The occurrence of radiation polymerization only above T_g of the matrix was also observed for organic glasses [15].

The cessation of polymerization before 100% conversion in vinyl polymerizations and curing reactions of thermosets has been known for a long time. In the case of vinyl polymerization, it was thought as due either to the occlusion of growing polymer radicals or to some unimolecular termination process.

Effects of glass transition have been discussed occasionally in a qualitative manner [4]. But a very clear quantitative explanation that polymerization practically stops when the overall polymerization system is vitrified was presented for the first time by the present authors. Kelley and Bueche [16] expressed the glass transition temperature T_g of a polymer-monomer (or plasticizer) mixture by Eq. (1), assuming the additivity of the free volumes of the two components:

$$T_g = \frac{\alpha_p \varphi_p T_{gp} + \alpha_m (1 - \varphi_p) T_{gm}}{\alpha_p \varphi_p + \alpha_m (1 - \varphi_p)} \tag{1}$$

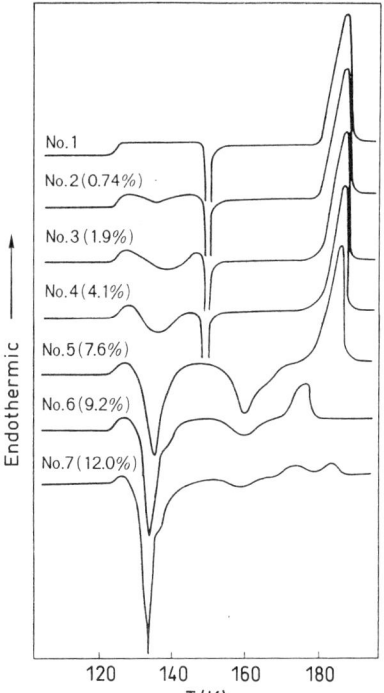

Fig. 1. DTA thermograms for unirradiated (No. 1) and irradiated (Nos. 2–7) vinyl acetate monomer glass [12,13]. $T_g = 122$ K, $T_{crystal} = 150$ K, and $T_{melt} = 180$ K. Figures in parentheses indicate the percent conversion to polymer

Here α is the difference in volume expansion coefficient of each component between the melt and glassy state, φ the volume fraction, and the subscripts p and m refer to polymer and monomer, respectively. The present authors [17] applied Eq. (1) with the proposal that the volume fraction φ_{p,T_g} at which viscous liquid transforms to glass can be calculated for a polymerization at temperature T from

$$\varphi_{p,T_g} = \frac{\alpha_m(T - T_{gm})}{\alpha_p T_{gp} - \alpha_m T_{gm} + (\alpha_m - \alpha_p)T} \qquad (2)$$

The calculated values of φ_{p,T_g} agreed well with the limiting conversion values for bulk polymerization of methyl methacrylate estimated by isothermal differential scanning calorimetry (DSC) [17] and gravimetry [18]. The theory can be applied not only to radical polymerization but also to anionic bulk polymerization of α-methylstyrene [19] and curing reactions of epoxides [20]. Thus, it was proved that many polymerization reactions stop when the system changes to the glassy state even before the complete conversion of monomer to polymer. This expectation is schematically shown in Fig. 2.

The photodimerization of cinnamic acid has long been believed to occur only in the crystalline state of appropriate geometry in which the distance between nearest-neighbor double bonds is smaller than 4 Å [21]. Thus it was considered, by the end of 1960's, that chemical reactions in amorphous polymers hardly ever occur below T_g. Recently, cyclizing imidization of aromatic poly(amic acid) [22] was observed to stop owing to vitrification before the complete conversion of amic acid to imide. The raising of the temperature provided additional progress of the reaction.

As is well known, molecular motions such as local mode relaxation of the main chain and rotational mode relaxation of the side chain are not frozen-in at T_g. The typical temperature dependence of viscoelastic properties is shown schematically in Fig. 3 [23] for both crystalline and amorphous polymers. Subglass transitions (T_β, T_γ, T_δ, etc.) corresponding to the freezing of these local motions appear as peaks in a tan δ-temperature curve.

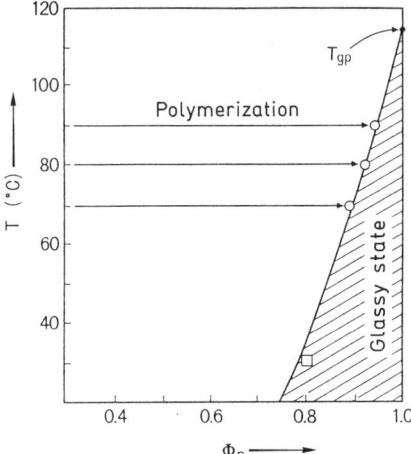

Fig. 2. Schematic diagram for the cessation of polymerization due to vitrification. (○, □): experimental results from PMMA bulk polymerization [17, 146]

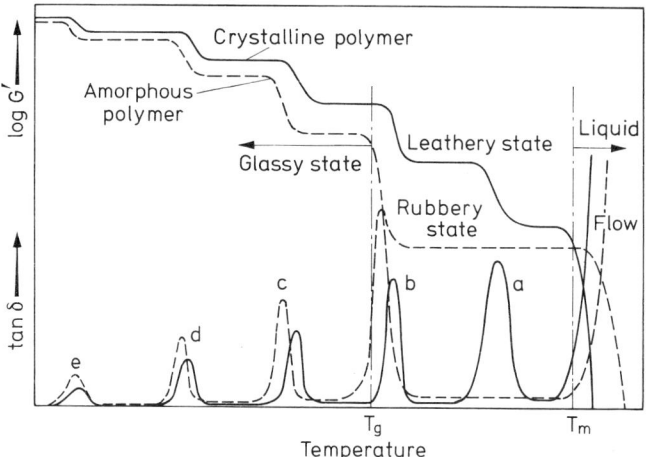

Fig. 3. Typical temperature dependence of dynamic modulus G' and loss factor tan δ for crystalline polymers (———) and amorphous polymers (- - - - -) [23]. (a) Chain motion in crystalline region, (b) Micro-Brownian motion of main chain, (c) Local motion of main chain, (d) Free rotation of side chain etc., (e) Rotation of α-methyl group etc.

Thus, reactions which need only local translation or rotation of reactants are supposed to proceed even below T_g; for example, photochemical chain scission of polymeric ketones by Hartley and Guillet [24], photochromic reactions of spirobenzopyran by Gardlund [25], photoisomerization of azocompounds by Kamogawa et al. [26], and photorearrangement of o-nitrobenzaldehyde by Cowell and Pitts [27]. Following those studies, the relationship between polymer mobility and chemical reactions in polymer solids above and below T_g have been systematically worked out, with the results discussed in the following sections. The exchange reactions of polycondensates in melt and in solid states as well as chemical healing at the interface or crack of bulk polymers have been studied at the level of chain dynamics and reviewed recently [28].

3 Characteristics of Reactions in Amorphous Polymer Solids

3.1 Factors Affecting Reactions in Polymer Solids

In contrast to topochemical reactions in the crystalline state where the crystalline structure and the distance between reacting chemical bonds are crucial [9, 10], the reactions in amorphous polymer solids are governed by the mobility and heterogeneity of reactive sites. Besides chemical reactivity, factors which affect the reactions in polymer solids are summarized in Table 1.

The mobility of reactants depends not only on the type of reaction as well as the size and shape of reacting groups, but also on the molecular motion and free volume distribution in the matrix polymer. Three types of heterogeneity can be distinguished. Microhetrogeneity due to a heterogeneous free volume distribution would lead to

Table 1. Factors affecting reactions in polymer solids

1. Mobility of reactants:
   a) Translational and rotational diffusion of reacting groups as functions of:
      i) Temperature and molecular motion of matrix. (T_g, T_p, T_γ, etc.);
      ii) size and shape of reacting groups;
      iii) free volume and conformation;
      iv) interaction between reacting groups and matrix.
   b) Cage effect.
2. Heterogeneity of the system:
   a) Macroheterogeneous system (gradient of absorbed light, oxygen concentration, etc.);
   b) macrohomogeneous but morphologically heterogeneous system (crystalline and amorphous parts, polymer blends);
   c) macro and morphologically homogeneous but microheterogeneous system (microheterogeneous distribution of reactive sites, free volume).
3. Change in matrix structure during reaction:
   a) More reactive sites are lost earlier;
   b) crosslinking, chain scission, change in absorbance, etc.
4. Factors other than mass transfer:
   a) Excited energy transfer and migration;
   b) electron transfer;
   c) transfer of active sites due to chain reaction.

heterogeneous progress of reaction even in a macroscopically and morphologically homogeneous system. Changes in matrix structure in the course of reaction, as observed in photocrosslinking of poly(vinyl cinnamate) [29], also lead to a change in reactivity. On the other hand, excited energy migration [30] or displacement of reactive sites by a successive chain reaction with adjacent monomer without mass transfer (reaction diffusion) [31, 32] facilitates the reactions in the solid state.

3.2 Effects of Molecular Motion and Temperature

We may expect that the mechanisms of solid-state reactions change with temperature in the following order:

1) Chemically-controlled stage: the reaction proceeds at the same rate as that in solutions (with normal activation energy).
2) Diffusion-controlled stage: the reaction proceeds heterogeneously, reflecting the molecular motion of the matrix polymer.
3) Freezing of the reaction due to the slowing of molecular motion.

In the case of the Norrish type II photolysis of ethylene-carbon monooxide copolymers [24] and vinyl ketone copolymers [33], the quantum yield in solid films above T_g was the same as that in solution, but it decreased with decreasing temperature T below T_g, and became almost negligible at T below T_β corresponding to the freezing of local mode relaxation of the main chain. The temperature dependence of the Norrish type II reaction is schematically shown in Fig. 4. Thus it is certain that the Norrish type II photolysis in polymer film obeys a chemically-controlled mechanism for $T > T_g$, a diffusion-controlled mechanism for $T_g > T > T_\beta$, and is frozen-in for $T < T_\beta$.

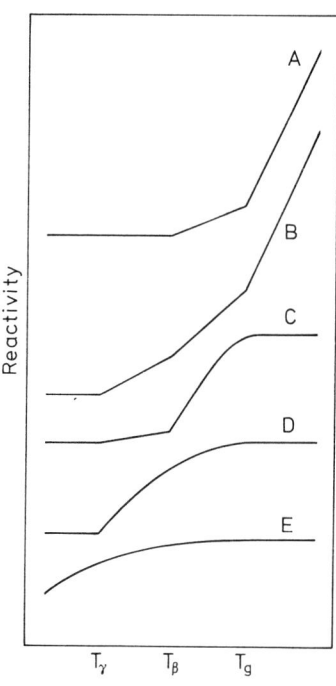

Fig. 4. Schematic temperature dependence of reactivity in solid polymers [40]. (A) Polymer-polymer reaction (diffusion controlled). (B) Polymer side chain-small molecule reaction (diffusion controlled). (C) Intramolecular main-chain scission. (D) Intramolecular reaction in polymer side chain. (E) Intramolecular reaction of small molecule with small critical free volume

When photochromic reactions of spirobenzopyran, azobenzene, and fulgide in polymer films [34] are compared, it is found that the critical free volume v_{fc} needed for a reaction is relatively large for spirobenzopyran, but very small for azobenzene and fulgide. The thermal decoloration reaction of spirobenzopyran in a solid film below T_g was much slower than in solutions [35] and was considered diffusion-controlled. Thermal cis to trans isomerization of azobenzene in polycarbonate at room temperature proceeded at a rate comparable to the rate in solutions, except for an anomaly at the initial stage [36–38]. Thus, the most part of this reaction can be considered chemically-controlled even below T_g of the matrix polymer.

Attempts have been made to relate the reaction rate in the solid state to the relaxation map of the matrix polymer [3, 39]. Rate coefficients for thermal backward reactions of photoisomerized spirobenzopyran and azobenzene are superimposed on the relaxation map of polystyrene, as shown in Fig. 5 [3]. The relationship $v_{f\beta} < v_{fc} < v_{fg}$ appears to hold for thermal decoloration of spirobenzopyran molecularly dispersed in a polymer film, where $v_{f\beta}$ and v_{fg} are the average free volumes at the β-relaxation temperature and T_g, respectively. The difference in reaction rate between film and solution appears in the diffusion-controlled region. In the molecularly dispersed azobenzene system, $v_{fc} < v_{f\beta}$ holds, and the rate coefficient intrinsic to the thermal cis → trans isomerization is much smaller than the frequency for the curve corresponding to the γ-relaxation of polystyrene. Thus, the thermal cis → trans isomerization of azobenzene in polymer film is chemically-controlled over a substantially entire temperature range. The relation between the diffusion-controlled decay of radicals produced by ^{60}Co γ-irradiation and the α-relaxation in polyethylene and polyoxymethylene was also discussed in terms of relaxation maps [39].

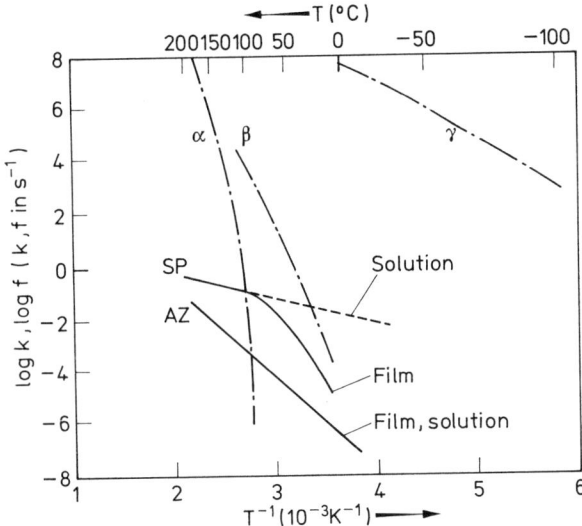

Fig. 5. Relationship between rate coefficients k for thermal isomerization in film (———) and in solution (------), and relaxation frequency f of matrix polymer (—·—·—). SP: merocyanine form of spiropyran, AZ: cis-azobenzene, matrix polymer: polystyrene

The temperature dependence of the rates or quantum yields of reactions in polymer solids is classified as shown in Fig. 4 [40], depending on the mode of molecular motion necessary for the reaction and also on the rate of molecular motion relative to the rate of intrinsic chemical reaction.

Case (A) corresponds to the polymer-polymer reactions controlled by microbrownian motion of the polymer main chain, which proceed very rapidly above T_g. Radiation degradation of polyisobutylene [41] and photodegradation of poly(vinyl acetate) [42] as well as photodimerization of polymer-bound anthryl groups [43,44] or cinnamonyl groups [45] show this type of temperature dependence.

Case (B) corresponds to the diffusion-controlled reactions of small molecules with functional groups on polymer side chains. Because the local mode of relaxation of a polymer main chain (β-dispersion) and the rotation of a polymer side chain or phenyl group in the main chain (γ-dispersion) occur below T_g, the reactions controlled by the side chain local motion begin at temperature T_γ corresponding to the onset of rotational motion and show breaks at T_β and T_g. Quenching or hydrogen abstraction of the benzophenone triplet in poly(methyl methacrylate) [46] may be explained in terms of this idea.

Case (C) corresponds to the photolysis of polymers containing methyl vinyl ketone [33] and methyl methacrylate [49], where the intramolecular main-chain scission is considered to determine the overall rate of photolytic reactions. At temperatures above T_g, the reactions are not at all affected by the mobility.

Case (D) was observed in the intramolecular photo-Fries reaction of the side chain ester group in poly(phenyl acrylate) [50].

Case (E) corresponds to the photoisomerization of azobenzene [38] or fulgide [34], whose critial free volume is small and some chromophore molecules are still in mobile sites below T_γ of matrix polymers and even at liquid helium temperature. The photodissociation of the carbon-oxygen bond in spirobenzopyran leading to the merocyanine form in polycarbonate film also proceeded rapidly even at temperatures below T_γ, e.g., at 80 K [51] and 4 K [34]. The disappearance of the temperature dependence characteristic of Cases (C), (D) and (E) at high temperatures can be attributed to the crossover of the reaction from diffusion controlled to chemically controlled, and the rates or quantum yields for these reactions at $T > T_g$ become equal to those in solutions. The temperature dependence of the diffusion of oxygen in polymer solids varied not significantly at T_g [52], but appreciably at T_β and T_γ. Thus the temperature dependence of the mobility of small molecules or reactive groups in polymer solids is substantially influenced by the size of the critical free volume necessary for diffusion or reaction to occur.

3.3 Heterogeneous Progress of Reactions

3.3.1 Deviation from First-Order Kinetics

The discussion in the previous section was concerned with the temperature dependence of average reactivity in polymer solids. However, as the reaction proceeds, the reactivity often changes owing to the non-homogeneous microenvironment around the reactants, leading to non-first-order reaction kinetics.

Two types of deviation from the first-order kinetics are noted for photo- and thermal isomerization reactions in polymer films. The first is the normal type, in which the reaction rate is the same as or smaller than that in solutions at the initial stage and then progressively becomes smaller. Typical examples are thermal decoloration of the photocolored merocyanine form of spirobenzopyran molecularly dispersed in or chemically bound to a polymer matrix [5, 25, 35] and photoisomerization of the transazobenzene residue incorporated in polymer main chains [36]. The first interpretation for the decoloration of the merocyanine form assumed the existence of different isomers, each of which fades independently following first-order kinetics [5, 53]. On the other hand, Kryszewski et al. [35] proposed the kinetic matrix effect, which means that the distribution of free volume may lead to the deviation from first-order kinetics. His idea was based on the finding that deviations from first-order kinetics can be observed even in simple molecules such as azobenzene which has only one trans or cis isomeric form. The effect of free volume distribution on reactivity was further demonstrated by studies of annealed polymer films [54, 55]. The distribution function of free volume as well as the critical free volume v_f were estimated for the merocyanine form of spiropyran in poly(methyl methacrylate) [35], derivatives of azobenzene in polystyrene [56], and azobenzene in polycarbonate [38]. The deviation from first-order kinetics was also observed in cyclizing imidization of model poly(amic acid) in a polyamide matrix [57].

The second type of deviation was the anomalous type, in which the rate of thermal cis → trans isomerization of azobenzene in polycarbonate film prepared by UV irradiation was initially faster than those in solutions and gradually approached the latter. Paik and Morawetz [36] attributed the anomalously fast matrix effect to the trapping of some cis-isomers in a strained conformation during UV-irradiation below T_g. On the other hand, Eisenbach [37] attributed this anomaly to the restricted chain relaxation and the fluctuation of free volume. The experiment with a cis-rich sample film cast under UV-irradiation [38] indicated a nearly first-order thermal isomerization. This finding supports the idea that residual strains in a cis-azobenzene film sample prepared by photoisomerization would be responsible for the anomalous deviation from first-order kinetics in polymer solids.

3.3.2 Photoreactions and Photophysical Processes

The time scale over which an event occurs determines what molecular motions and interactions influence a particular event. In the case of photoisomerization, the reaction proceeds only with the molecules which lie at the sites where the reactant is mobile during the excited lifetime. Consequently a very small quantum yield or an apparent rate coefficient is often observed for the photoreaction which requires a large critical free volume, e.g. for the photodecoloration of the merocyanine form of spirobenzopyran in polycarbonate film [51].

In a photophysical process with some critical free volume, e.g., excimer formation of oligostyrene diastereomers in PMMA [58] and 1,3-di-(1-pyrenyl)propane in polystyrene [59], the steady-state fluorescence intensity ratio of the excimer to the monomer may give the fraction of preformed excimer sites or mobile sites during the excited lifetime. In contrast, such photochemical reactions as photoisomerization and photodimerization under stationary-state light irradiation accumulate the reaction product so that their progress can be detected even when the quantum yield is very small [51, 60]. The cumulative nature of photoisomerization and photodimerization in a mobile system is associated with the redistribution of free volumes which allows a large extent of reaction to proceed. This is the case even though the instantaneous fraction of a chromophore which has a local free volume larger than the critical free volume is very small. For a fluorescence probe, the excimer formation is not cumulative, but has to do only with chromophores which can move during the excited lifetime of the nanosecond to submicrosecond time scale.

The redistribution of free volumes also influences the sub-glass transition temperatures T_β and T_γ observed for photoisomerization reactions in polymer solids. T_g, T_β and T_γ are frequency-dependent, and the response of any process to the transitions at these temperatures depends on the time scale. The time scale of photoprocesses may not be equal to those of DSC or dynamic mechanical methods, which are of the order of 10^{-1} to 10^3 Hz. However, for photodecoloration of the merocyanine form of spirobenzopyran in polycarbonate film under steady-state irradiation of 560 nm light after laser-single-pulse induced coloration [51], it was found that the Arrhenius plot of the apparent rate coefficient broke at T_g (150 °C), T_β (20 °C), and T_γ (−120 °C) of the matrix polycarbonate; these temperatures are the ones determined by dynamic mechanical measurements. The excited state lifetime of the merocyanine form in polycarbonate was 1.8 ns [61]. Hence, the decolorating isomerization during the lifetime proceeded only in a small fraction of the molecules surrounded by a sufficient amount of free volume. Thus, it is likely that the temperature dependence of the apparent rate coefficient reflecting the relative quantum yield is controlled by the frequency of redistribution of free volumes, which may be comparable with the frequency determined by dynamic mechanical measurements.

3.3.3 Effects of Binding to the Polymer Chain on the Reactivity of Chromophore

The covalent binding of a chromophore to the polymer main chain restricts its mobility and increases its critical free volume v_{fc}. Therefore, it should decrease the reaction rates and should deviate unimolecular reactions more appreciably from first-order kinetics.

Smet et al. [62] investigated the effect of the molecular size of the spirobenzopyran chromophore on reactivity due to the binding of the indoline and benzopyran rings with oligo- and polyester chains. They found a marked slowdown of the thermal decoloration rate of the chromophore with the increase in molecular size, when its color was induced by UV irradiation above T_g of the matrix polyester. Though the stretching of the polymer matrix containing spirobenzopyran groups also retarded the rate of thermal decoloration [5], the influence of increasing molecular size on the slowdown of isomerization was overwhelming.

Morawetz and Paik-Sung prepared poly(methyl methacrylate) (PMMA-Az (s)) [36] and polystyrene (PS-Az (s)) [56] containing azo-chromophore in their side chains, and polyamide(Nylon-Az (m)) [36], polyester(PES-Az (m)) [36], polyurethane (PU-Az (m)) [63], and polystyrene (PS-Az (m)) [56] containing the same chromophore in their main chains, and compared the photoisomerization reactivities of these compounds in bulk with those in solution. Table 2 summarizes their experimental results, where x_∞ denotes the final or equilibrium extent of reaction of the chromophore, together with the corresponding data for azobenzene molecularly dispersed in a polymer matrix [38]. The values of x_∞ are comparable in bulk and in dilute solution both for molecularly dispersed azobenzene and for azobenzene residues in the side chain irradiated above T_g. The bulk samples with azo-chromophore in the side chain irradiated below T_g yielded a substantially lower conversion than that in solution. Thus, incorporation of azobenzene groups in the polymer main chain by binding two phenyls of the azo groups drastically inhibits photoisomerization in bulk, especially when irradiated below T_g. A similar tendency was observed for the apparent rate or quantum yield of photoisomerization of azochromophores. The apparent rate for a molecularly dispersed sample are comparable in bulk and in solution at room temperature [38]. The

Table 2. Final or equilibrium extent of reaction x_∞ for trans→cis photoisomerization of azobenzene group molecularly dispersed, bound to side chain (s), chain end (e), or incorporated in main chain (m) of various polymers

Chromophore	x_∞		in	Ref.
	Solution	Film (T > T_g)	Film (T < T_g)	
Azobenzene	0.90		0.90	[38]
PMMA-Az (s)	0.76	0.76	0.54	[36]
PEMA-Az (s)	0.78	0.78	0.61	[37]
PS-Az (s)	0.67		0.30	[56]
PS-Az (e)	0.85		0.33	[56]
PS-Az (m)	0.74		0.06	[56]
PU-Az (m)	0.80	0.62	0.39–0.47	[63]
Nylon-Az (m)	0.60		0.12	[36]
PES-Az (m)	0.45	0.18	0.03	[36]

relative quantum yield for the side-chain bound chromophore in bulk below T_g was considered to be about one fifth of that in solution [36]. From the microscopic point of view, it is pertinent to consider that the distribution of reactivity is associated with the heterogeneity of reactive sites. In analyzing the trans → cis photoisomerization of the azobenzene chromophore attached to the chain end, chain center, or side group of the polystyrene chain [56], only some fraction of the chromophore was assumed to isomerize at the same rate as in dilute solution. In the next section we will refer to some recent studies which deal with solid state reactions as a dispersive process affected by the free volume distribution or site-energy distribution.

3.4 Kinetic Analysis of Non-Homogeneous Reactions

In general, chemical reactions and random physical processes taking place in solid media such as amorphous polymers follow non-exponential kinetics even if there is good reason to expect that the elementary step be a first-order process. The non-exponential decoloration rate of the merocyanine form of spirobenzopyran was explained by Kryszewski et al. [35] in terms of the distribution of free volumes in the matrix. By assuming the equilibrium to exist between the expanded and collapsed forms (B_{expand} and $B_{collapsed}$) of microenvironments around the merocyanine (B) and the thermal decoloration to occur only from the expanded form, they succeeded in deriving the experimentally observed $\sqrt{t}$ dependence of the decoloration reactions.

$$B_{collapsed} \rightleftarrows B_{expanded} \xrightarrow{decoloration} A \quad (3)$$

Azobenzene residues in a polymer chain photoisomerize by a single rate process in dilute solution, but the rate of this reaction in solid films consists of two processes. One is as fast as in dilute solution and the other is slower. The fractional contribution of the former decreases with physical aging, but increases with temperature, plasticization, or tensile deformation [54, 55]. Sung et al. [56] divided the microenvironments of the azobenzene residue into freely mobile and frozen sites, and related the fraction α of the former to the area in the free volume distribution profile in which the local free volumes are larger than the critical size for photoisomerization. By so doing, the trans content A as a function of time t during the photoisomerization was shown to be given by

$$A = (1 - \alpha) + \alpha\{A_e + (1 - A_e) \exp[-(k + k')t]\} \quad (4)$$

where A_e is the trans content in the apparent photostationary state, and $k + k'$ the sum of forward (trans → cis) and backward (cis → trans) reaction rate coefficients, which may be evaluated from dilute solution data.

The temperature dependence of photoisomerization in polymer solids yields detailed information about the effects of free volume distribution and molecular motion on the reaction process. The present authors [38] measured the rate of trans → cis photoisomerization of azobenzene in a polycarbonate film over a very wide range of temperatures (4 K–423 K), and proposed a kinetic model, which consists of chemically controlled and diffusion-controlled reactive sites, by considering the free volume distribution and the critical free volume f_c. The extent of reaction, $R(t)$, can be expressed by an integration of uncoupled individual first-order reactions with the rate coefficient $k(f)$ for a fractional free volume f and $G(f)$, the distribution function of f, as follows:

$$R(t) = \int_0^\infty G(f) \exp[-k(f)t] \, df \qquad (5)$$

The rate coefficient $k(f)$ is equal to k_0, the apparent rate coefficient for photoisomerization in solution, for $f > f_c$.

The experimental data for the temporal decay of a reacting species embedded in a random matrix can often be approximately described by a stretched exponential [65, 66]

$$\varphi(t) = \exp[-(t/t_0)^\alpha] \qquad (6)$$

where α measures the deviation from first-order kinetics and t_0 characterizes the time scale of the reaction. Equation (6) has been correlated to a sequence of relaxation processes [67], in which faster processes are followed by slower ones. For example, the relaxation of a dipole embedded in a polymer matrix, if triggerd by time dependent diffusion of a defect that provides the free volume required for molecular motion, would obey this mechanism [68-70]. However, it has been shown [71, 72] that the validity of Eq. (6) is no longer restricted to serial processes. Tsutsui et al. [73] applied this equation to analyzing thermal decoloration of photochromic spiroindolinonaphthoxazine in glassy polymer matrices. The dispersion parameter α was found to be unity above T_g and to become close to 0.5 below T_g for the samples colored by continuous UV irradiation, while α approached 0.3 below T_g for the samples colored by laser pulse irradiation.

Richert et al. [66, 74, 75] compared experimental results on thermal decoloration of the merocyanine form of spirobenzopyran in a PMMA matrix with Eq. (6) as well as Eq. (5). The stretched exponential with $\alpha = 0.366$ provided a reasonably good agreement with the experimental data if applied to a limited time interval. However, analysis of the short time data indicated the failure of Eq. (6) for modelling decoloration. Instead, uncoupled parallel relaxations in which the rate controlling parameter, an activation energy, was given by a Gaussian distribution made an appropriate description of the observed decoloration profile possible. Using the stretched exponential and the uncoupled parallel relaxation, Richert et al. [66] also examined the rate of photophysical hole formation in the absorption band of tetracene in a methyltetrahydrofuran glass at low temperature and the diffision-controlled trapping of a triplet excitation in a benzophenone glass.

4 Photophysical Processes

In what follows, we discuss the non-exponential decay of phosphorescence in polymer matrices and its origin, along with the effect of a multi-photon process on decay curves. Singlet and triplet energy migrations and excimer formation [6, 76] are important photophysical processes in polymer solids as they are in solution, and will be discussed separately.

4.1 Non-Exponential Phosphorescence Decay in Polymer Solids

The general photophysical processes originated from the excited triplet state includes phosphorescence (k_{PT}), nonradiative deactivation (k_{TT}), triplet energy transfer and migration, and T-T annihilation [77, 78]. The triplet-triplet energy transfer from donor A to acceptor B was first clearly demonstrated by Terenin and Ermolaev [79, 80]. The triplet energy transfer needs the overlap of electron clouds of the donor and the acceptor, and its efficiency depends on the gap E_T between the energy levels of the excited triplet states of the donor and the acceptor. Such an energy transfer by electron exchange interaction is called the Dexter mechanism after Dexter [81] who theoretically formulated it.

When chromophores are sufficiently mobile as in fluid solutions, the bimolecular quenching process which includes the triplet energy transfer and some collisional quenching does not alter the single exponential nature of phosphorescence decay of donor chromophore A.

$$^3A^* + B \rightarrow A + B \quad \text{triplet quenching } (k_q) \tag{7}$$

In other words, the phosphorescence decay profile I(t) may be expressed by

$$I(t) = \exp(-t/\tau) = \exp\{-t[1/\tau_0 + k_q[B]]\} \quad \text{(Stern-Volmer)} \tag{8}$$

with a single lifetime τ, which is related to the phosphorescence lifetime in the absence of quencher, $\tau_0 = 1/(k_{PT} + k_{TT})$, the quenching rate constant k_q and the quencher concentration [B] by Eq. (9) known as the Stern-Volmer relationship [82].

$$1/\tau = 1/\tau_0 + k_q[B] \tag{9}$$

As an extreme contrast Perrin [83] considered the case in which the donor and acceptor molecules are immobile and the energy transfer occurs instantaneously only when the two molecules come close to a critical transfer distance R_0. His formulation gives the following decay function of donor phosphorescence:

$$I(t) = \begin{cases} 1 & (t = 0) \\ \exp[-(t/\tau_0) - (C_B/C_0)] & (t > 0) \end{cases} \quad \text{(Perrin)} \tag{10}$$

where C_B is the acceptor concentration and C_0 the critical transfer concentration which is defined as the reciprocal volume of the sphere with radius R_0.

The Perrin model oversimplifies the reality. The static triplet-triplet energy transfer between immobile chromophores dispersed in solids can be better described by the Inokuti-Hirayama theory [84] (Eq. (11)) based on the Dexter mechanism with a distance-dependent rate coefficient for triplet energy transfer.

$$I(t) = \exp[-t/\tau_0) - \gamma^{-3}(C_B/C_0) G(e^\gamma t/\tau_0)] \quad \text{(Inokuti-Hirayama)} \tag{11}$$

where $\gamma = 2R_0/L$ with L being the Debye radius and G(Z) a function of Z. Mataga et al. [85] were the first to study accurately the energy transfer from the excited triplet of benzophenone to naphthalene by laser flash photolysis at 77 K. They showed that the non-exponential decay curves of the benzophenone triplet obey the Inokuti-Hirayama equation (Eq. (11)).

The Inokuti-Hirayama theory predicts a single-exponential decay for the phosphorescence of a chromophore in the absence of an acceptor. However, as is well known, the phosphorescence decay of organic molecules molecularly dispersed in polymer matrices sometimes is not single-exponential even for the samples containing no other additives. Hence, another mechanism must be sought for the non-exponential decay of chromophores observed in such polymer matrices.

Non-exponential phosphorescence decay of some chromophores dissolved in plastics were first noted by Oster et al. [86]. Such decay curves were found for naphthalene and triphenylene phosphorescence in poly(methyl methacrylate) (PMMA) at room temperature [87]. However, the decay in PMMA at room temperature were exponential for the anthracene triplet [88], pyrene [89], and coronene phosphorescence [87]. Graves et al. [90] analyzed the temperature dependence of the phosphorescence parameters for a number of aromatic hydrocarbons in PMMA at 77–400 K and suggested an intermolecular thermally assisted energy transfer to occur from the chromophores to the host plastic at higher temperature.

El-Sayed et al. [87] interpreted the non-exponential decays of naphthalene and triphenylene in PMMA at room temperature in terms of a triplet-triplet annihilation mechanism. The basis of their discussion was the decrease in non-exponentiality with the decrease in excitation intensity and the observation of delayed fluorescence. El-Sayed et al. suggested that a non-exponential decay should be observed for molecules with first-order lifetimes of the order of several seconds. However, coronene with the triplet lifetime of 8.5 s gave an exponential decay. These authors also suggested a rather high concentration of chromophore at the excited triplet state on the basis of the diffusion-controlled triplet-triplet annihilation mechanism. Later, Jassim et al. [91] proposed another type of triplet-triplet annihilation mechanism for the non-exponential phosphorescence decay. It consisted of the energy transfer from the phosphorophore to the matrix polymer, the triplet energy migration through the matrix polymer, and the triplet-triplet annihilation between the chromophore triplet and the polymer triplet.

Horie and Mita [92] measured the phosphorescence decay of benzophenone in PMMA over a wide temperature range (80–433 K). Non-exponential decays were observed between T_β (onset of ester side group rotation of the matrix polymer) and T_g (glass transition temperature). The decay profile was independent of the intensity of excitation laser pulse over a 100 times intensity change [93]. Thus, Horie and Mita attributed the observed non-exponential decays to a single photon process including the intermolecular dynamic quenching of the benzophenone triplet by the ester groups in the side chain of PMMA. A detailed discussion of the dynamic quenching mechanism will be given in Sect. 4.3.

It is noteworthy that reported triplet decay curves of various chromophores in PMMA at room temperature can be divided approximately into two groups, according to E_T of the chromophores, as shown in Table 3. Non-single-exponential phosphorescence decay were observed for chromophores with E_T not much smaller than

Table 3. Types of triplet decay curves and triplet energies E_T for various chromophores dispersed in PMMA at room temperature

Chromophore	Type of triplet decay	E_T [95] kJ/mol	Refs.
Fluorene	Non-single-exponential	284	[87]
Benzophenone	Non-single-exponential	283	[46, 92]
Triphenylene	Non-single-exponential	278	[87, 91]
Phenanthrene	Single-exponential	260	[91]
Naphthalene	Non-single-exponential	255	[87]
	Single-exponential		[91]
Coronene	Single-exponential	228	[87]
Benzil	Single-exponential	227	[94]
Pyrene	Single-exponential	202	[89]
Anthracene	Single-exponential	179	[88]

that of PMMA (297–301 kJ) [90, 91]. On the other hand, single-exponential phosphorescence or T-T absorption decays were obtained for chromophores with lower triplet energies. These results also support the idea that dynamic quenching of excited-triplet-state chromophores by matrix PMMA takes place as a result of the endothermic triplet energy transfer.

4.2 Phosphorescence Decay of Benzophenone in Polymer Solids

Typical decay curves for benzophenone (BP) phosphorescence at 450 nm in PMMA excited by a 10-ns nitrogen laser pulse at 337 nm are shown as a function of temperature in Fig. 6 [46]. The phosphorescence intensity I(t) decreases single exponentially below the temperature corresponding to the onset of ester side-group rotation ($T_\beta = -30$ °C for PMMA). The deviation from a single-exponential decay, observed for $T > T_\beta$, is enhanced with increasing temperature, but it becomes less marked above T_g of the matrix polymer and disappears at 150 °C. A similar temperature dependence of the decay profile was observed for benzophenone phosphorescence in other acrylic polymers [96], polystyrene (PS), and polycarbonate (PC) [47].

Transient spectra showed that these benzophenone phosphorescence decays were due only to the benzophenone triplet [46], and the irradiation intensity independence ruled out the T-T annihilation under the experimental conditions used [93]. Since the deviation was not observed at temperatures below T_β of each acrylic polymer or T_γ of polystyrene and polycarbonate, this quenching is considered not of the static character as featured in the Inokuti-Hirayama theory, but of a dynamic type associated with the collision of functional groups and the energy transfer with some activation energy.

Annealing above T_g and slow cooling were found to affect the benzophenone phosphorescence decay in PMMA [92]. The occurrence of quenching was ascertained by the rate constant for benzophenone triplet quenched by methyl acetate in an aceto-

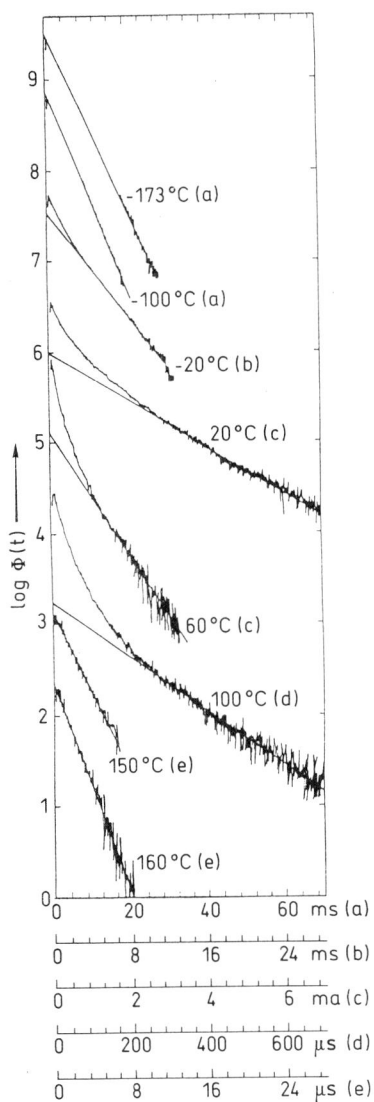

Fig. 6. Semilogarithmic decay curves of benzophenone phosphorescence in PMMA excited by 10-ns nitrogen laser pulse at 337 nm. Temperature and symbols for time scales are given beside the curves [46]

nitrile solution (3.9×10^3 M^{-1} s^{-1} at 30 °C) [96]. Benzophenone triplet quenching by the phenyl or phenylene group in polystyrene or polycarbonate is also known (1.2×10^6 M^{-1} s^{-1} for polystyrene in benzene at 30 °C) [97]. These quenching rate constants in nonviscous solutions are of a reasonable order of magnitude for an uphill-type endothermic triplet energy transfer.

Since the dynamic quenching of a phosphorophore gives a single exponential decay in nonviscous solutions (Stern-Volmer Model), it is of interest to ask why the dynamic quenching in polymer solids, especially below T_g, has a non-exponential decay profile.

4.3 Kinetics for Non-Exponential Decay Due to Dynamic Quenching

The decay of triplet benzophenone ^{3}BP* follows the processes:

$$^3B^* \xrightarrow{k_0} BP \tag{12}$$

$$^3BP^* + [Q] \xrightarrow{k_q} BP + [Q] \tag{13}$$

where $k_0 = k_{PT} + k_{IP}$ is the rate constant for spontaneous deactivation of the benzophenone triplet, and [Q] the concentration of ester, phenyl, or phenylene groups in the matrix polymer. The bimolecular rate coefficient k_q is given by Eq. (14) including both diffusion and chemical steps [98]

$$k_q = \frac{4\pi RDN}{1 + 4\pi RDN/k}\left[1 + \frac{R}{(1 + 4\pi RDN/k)\sqrt{\pi DT}}\right] \tag{14}$$

where D is the sum of the diffusion coefficients for the carbonyl groups in benzophenone and for the quenching groups in the polymer, R the reaction radius between the two groups, k the intrinsic (chemical) rate constant that would be obtained if the equilibrium concentration of the quenching groups were maintained, and N the Avogadro constant divided by 10^3. When k_q is controlled by diffusion of the two groups (i. e. $k \gg 4\pi RDN$), Eq. (14) reduces to

$$k_q = 4\pi RDN(1 + R/\sqrt{\pi Dt}) = A + B/\sqrt{t} \tag{15}$$

with

$$A = 4\pi RDN, \quad B = 4R^2(\pi D)^{1/2} N$$

Thus, the rate coefficient k_q includes a time-dependent term that dominates the very early stage of reaction where the steady-state diffusion of the quenching groups is not yet attained.

The decay rate of the benzophenone triplet is expressed by

$$\begin{aligned}-d[^3BP^*]/dt &= (k_0 + k_q[Q])\,[^3BP^*] \\ &= (k_0 + A[Q] + B[Q]\,t^{-1/2})\,[^3BP^*]\end{aligned} \tag{16}$$

from which we obtain

$$[^3BP^*] = [^3BP^*]_0 \exp\left[-(k_0 + A[Q])\,t - 2B[Q]\,t^{1/2}\right] \tag{17}$$

for the concentration of the benzophenone triplet $[^3BP^*]$ as a function of time, with $[^3BP^*]_0$ being $[^3BP^*]$ at $t = 0$. Since the phosphorescence intensity I(t) is proportional to $k_{PT}[^3BP^*]$, Eq. (16) yields

$$\begin{aligned}\ln I(t) &= -(k_0 + A[Q])\,t - 2B[Q]\,t^{1/2} \\ &= -(t/\tau) - C(t/\tau)^{1/2}\end{aligned} \tag{18}$$

where

$$1/\tau = k_0 + A[Q] = k_0 + 4\pi RDN[Q] \quad (19)$$

$$B = C\tau^{-1/2}/(2[Q]) = 4R^2(\pi D)^{1/2} N \quad (20)$$

Fitting Eq. (18) to an observed phosphorescence decay curve allows the reciprocal lifetime $1/\tau = k_0 + A[Q]$ and the parameter B to be estimated.

Figure 7 illustrates this method applied to PMMA and polystyrene. The breaks are observed at T_g and other subglass transition temperatures. The breaks at $T_{\alpha'}$ (PMMA) and T_β (PS) indicate the changes in local mode relaxation of the main chain, and those at T_β (PMMA) and T_γ (PS) the onset of rotation of the side-chain ester or phenyl group.

The Arrhenius plot of $1/\tau$ for benzophenone in poly(methyl acrylate) (PMA) [46] showed another break at 40 °C (above T_g of PMA), which corresponds to the crossover of k_q given by Eq. (14) from a diffusion-controlled to an activation-controlled reaction. The diffusion coefficient D for reacting carbonyl groups calculated from the values of $1/\tau$ and B also showed a break at each transition temperature, as exemplified in Fig. 8 for PMMA, polystyrene, and polycarbonate. It should be noted that D in Fig. 8 refers to the reacting functional groups but not to the molecule. The diffusion process at temperatures below T_g would be caused by rotation of the benzophenone molecule and by the cooperative motion of a few successive monomer units of the matrix polymer. Nevertheless, the values of D in these polymers at 100 °C are comparable to the value of $D = 5.6 \times 10^{-13}$ cm^2/s [99] for mass diffusion of ethylbenzene in polystyrene at 30 °C. The reaction radius R was estimated to be 3–5 Å. The transition temperatures

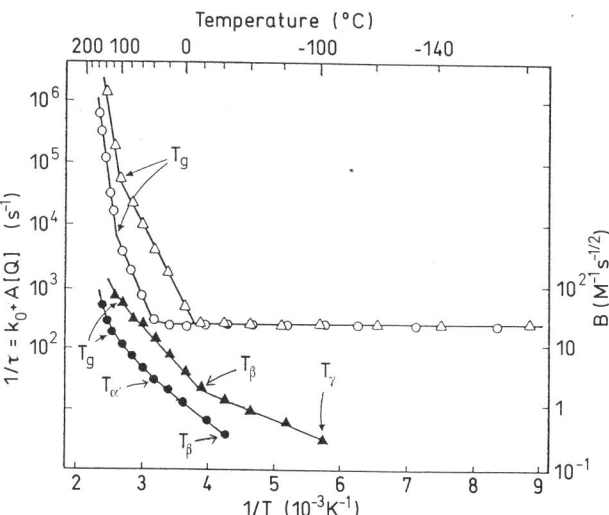

Fig. 7. Temperature dependence of reciprocal lifetime $1/\tau$ (○, △) and contribution of non-exponential term B (●, ▲) for benzophenone phosphorescence in PMMA (○, ●) and in polystyrene (△, ▲)

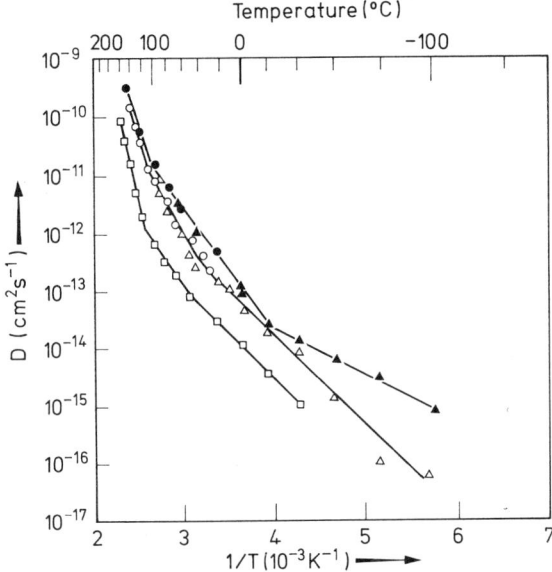

Fig. 8. Arrhenius plots for diffusion coefficient D of interacting groups in dynamic quenching of benzophenone triplet by phenyl, phenylene, and ester groups in polystyrene (●, ▲), polycarbonate (○, △), and PMMA (□)

of the matrix polymers monitored by phosphorescence decay of benzophenone are summarized in Table 4. The α' transition for PMMA and other acrylic polymers and the β transition for polystyrene, polycarbonate, and poly(vinyl alcohol) can be attributed to the local mode relaxation of the main chain. The phosphorescence probe technique is effective for the detection of this mode of sub-glass transition as well as other rotational modes of transition (T_β and T_γ) of polymer matrices.

Table 4. Transition temperatures of matrix polymers monitored by benzophenone phosphorescence and quenching reaction radius R

	T_β (°C)	$T_{\alpha'}$ (°C)	T_g (°C)	R (Å)	Refs.
Poly(methyl methacrylate)	−40	40	110	5.0	46)
Poly(isopropyl methacrylate)	−70	20	80	5.3	46)
Poly(methyl acrylate)	−70		10	(3.0)	46)

	T_γ (°C)	T_β (°C)	T_g (°C)	R (Å)	Refs.
Polystyrene	−100	−20	100	5.8	47)
Polycarbonate	−100	20, 100	150	5.2	47)
Poly(vinyl alcohol)	−100	30	85	3.5	48)

4.4 Hydrogen Abstraction of Benzophenone Triplet in Poly(vinyl alcohol)

The phosphorescence of benzophenone in a poly(vinyl alcohol) (PVA) film excited by a 10-ns nitrogen laser pulse at 337 nm decayed exponentially for $T < T_\gamma$ (-100 °C) or $T > T_g$ (85 °C), but nonexponentially for $T_\gamma < T < T_g$ [48]. The latter behavior was attributed to the diffusion-controlled hydrogen abstraction of the benzophenone triplet from the poly(vinyl alcohol) matrix. The occurrence of a non-exponential decay in poly(vinyl alcohol), in which no triplet energy migration is possible, gives additional evidence for the absence of the T-T annihilation mechanism in this case, because the T-T annihilation should occur through the triplet energy migration process [91].

The diffusion-controlled rate coefficient k_a for the above-mentioned hydrogen abstraction and the phosphorescence intensity $I(t)$ are given, as in the physical quenching of benzophenone phosphorescence by matrix polymers, by the relations

$$k_a \cong 4\pi RDN(1 + R/\sqrt{\pi Dt}) = A + B/\sqrt{t} \tag{21}$$

$$\ln I(t) = -(k_0 + A[PVA]) t - 2B[PVA] t^{1/2} \tag{22}$$

The fitting of Eq. (22) to an experimental phosphorescence decay curve gives the values of $1/\tau = k_0 + A[PVA]$ and B. The breaks observed at $T_g = 85$ °C and $T_\beta = 30$ °C and the appearance of B at $T_\gamma = -100$ °C clearly reflected the change in molecular motion of the matrix poly(vinyl alcohol). The quantum yields for hydrogen abstraction or ketyl radical formation f_a can be calculated from k_0 and A for each temperature. They are nearly unity.

In order to find the net quantum yield $\Phi(-BP)$ for the benzophenone disappearance by hydrogen abstraction in poly(vinyl alcohol), the UV spectrum of benzophenone at 256 nm in a poly(vinyl alcohol) film was measured under continuous irradiation of 365 nm UV light. The $\Phi(-BP)$ given in Table 5 is very small for $T < T_g$ as compared with f_a, suggesting the backward reaction (k_{back}) of the benzophenone ketyl radical to dominantly occur in the cage at temperatures below T_g. The T-T absorption spectrum and the lifetime of the ketyl radical were observed at $T > 120$ °C [48]. Figure 9 shows a plausible reaction scheme for the photochemistry of benzophenone in poly(vinyl alcohol) including the cage backward reaction.

Table 5. Quantum yield $\Phi(-BP)$ for benzophenone disappearance during 365 nm irradiation and fraction f_a of ketyl radical formation from transient measurements [48]

Temperature (°C)	I_0 (einstein/cm² · s)	$\Phi(-BP)$	f_a
0	2.6×10^{-8}	0.018	0.85
30	1.9×10^{-8}	0.042	0.95
	1.4×10^{-8}	0.056	
	8.1×10^{-9}	0.049	
	8.0×10^{-9}	0.052	
60	2.6×10^{-8}	0.075	0.99
100	1.9×10^{-8}	0.32	1.0
140	1.9×10^{-8}	0.49	1.0

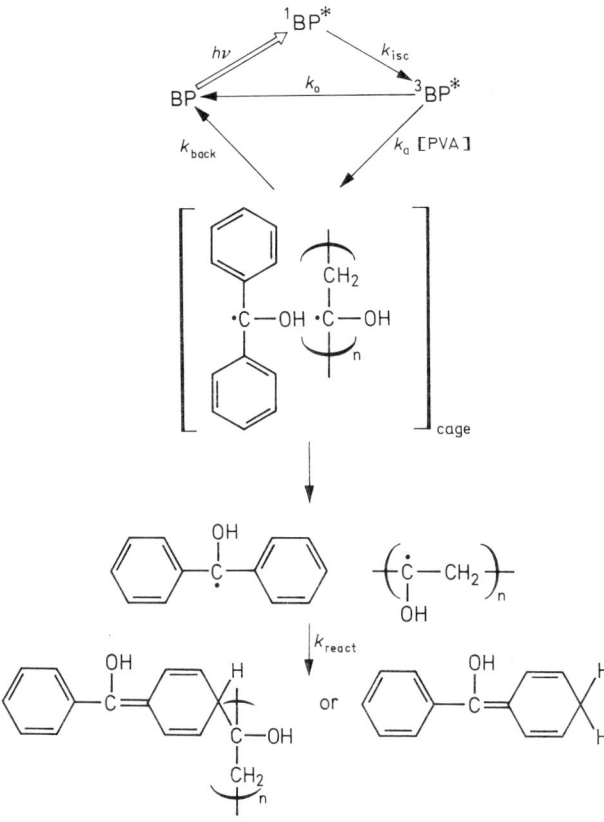

Fig. 9. Reaction scheme of photo-excited benzophenone in poly(vinyl alcohol) [48]

4.5 Phosphorescence Decay of Benzophenone under Multi-Photon Conditions

Salmassi and Schnabel [100] measured the decays of phosphorescence and the triplet absorptions of benzophenone in PMMA and polystyrene under high intensity irradiation of 347 nm frequency-doubled ruby laser single pulse. The initial triplet concentration was as high as 6×10^{-4} mol/l, in comparison with that of less than 6×10^{-6} mol/l in the previous case [46, 93] studied with nitrogen laser pulse. For PMMA, a single triplet decay mode following first-order kinetics was observed at T < 150 K and at T > 410 K, and two distinctly different triplet decay modes appeared in between. The latter consists of a fast first-order process, with a lifetime (ca. 4 μs at 295 K) independent of the initial triplet concentration $[T]_0$, and a slow second-order process with a first half-lifetime proportional to $[T]_0^{-1}$. Similar modes of triplet decay were also obtained with polystyrene matrices for 180 K < T < 350 K. Thus, Salmassi and Schnabel [100] concluded that the triplet-triplet (T-T) annihilation is an important deactivation process in the temperature range between T_g and T_γ, where the rotation of the α-methyl group (PMMA) or the phenyl groups (polystyrene) commences.

The rate of decay in triplet concentration in the presence of the T-T annihilation mechanism is given by

$$-d[T]/dt = \Sigma\, k_1[T] + k_{TT}[T]^2 \tag{23}$$

where $\Sigma\, k_1$ denontes the sum of all first-order or pseudo first-order rate constants of the triplet decay, and the T-T annihilation dominates when $\Sigma\, k_1 \ll k_{TT}[T]$. Salmassi and Schnabel[100] attributed the "fast" decaying triplets to triplets formed in close proximity, while they considered that the "slow" decaying triplets are formed at positions separated by relatively large distances and require a diffusive process for their interaction. Using the value of $k_{TT} = 5 \times 10^7\ M^{-1}\ s^{-1}$ at 295 K in PMMA and the Smoluchowski equation for the diffusion controlled T-T annihilation rate constant k_{TT}, they obtained a value of about $1 \times 10^{-8}\ cm^2/s$ for the diffusion coefficient of benzophenone in PMMA. This value is comparable in the order of magnitude with D of oxygen in PMMA, but is much larger than $D = 5.6 \times 10^{-13}\ cm^2/s$ [99] for ethylbenzene in polystyrene at 100 °C. It is likely that the triplet energy migration through the matrix polymer causes T-T annihilation between the chromophore triplet and that the polymer triplet (24) is necessary for the T-T annihilation to take place in polymer matrices below T_g.

Salmassi and Schnabel[100] also noted that hydrogen abstraction by the benzophenone triplet from PMMA and polystyrene became more appreciable at higher temperatures. The quantum yield for the ketyl radical formation for benzophenone in PMMA at 430 K was estimated to be $\Phi(BPH) = 0.06$. This value implies that only about 12% of the benzophenone triplet converted to the ketyl radical. Hydrogen abstraction by the benzophenone triplet from PMMA may proceed mainly through the two-photon process[46]. The fraction of ketyl radical formation was about 20% in the polystyrene matrix.

The non-exponential decay of benzophenone phosphorescence in PMMA at room temperature was also observed by Fraser et al.[101] under repeated irradiation of nitrogen laser pulse. These authors proposed a triplet-triplet annihilation mechanism in which the polymer matrix itself plays a role as an energy acceptor from the benzophenone triplet and also as a medium for triplet energy migration.

We have reviewed various mechanisms proposed to explain observed non-exponential phosphorescence decays of aromatic chromophores in rigid glasses and polymer solids. In the presence of an effective triplet energy acceptor, the static Dexter-type energy transfer to the acceptor molecule leads to the Inokuti-Hirayama type non-exponential triplet decay (Eq. (11)). The non-exponential decay of phosphorophores containing no other additives in polymer solids can be attributed to dynamic quenching by polymer matrices with the diffusion-controlled rate coefficient including the time-dependent transient term (Eq. (14)). The non-exponential phosphorescence decay under high-intensity and/or repeated laser pulse irradiation can be explained in terms of the biphotonic triplet-triplet annihilation mechanism, which is probably associated with triplet energy migration through polymer matrices.

ns and Photodynamics in Polymer Solids

5 Isomerization Reactions

As has been outlined in Chapt. 3, the isomerization reactions in amorphous polymer solids are appreciably influenced by local mobility and heterogeneity of reactive sites, often leading to the deviation of reaction profiles from first-order kinetics. However, this situation allows us to obtain an insight into the microstructure of amorphous polymer solids, e.g. distribution of local free volume, by using photoisomerization reactions as molecular probes. Since the photochromic phenomena in polymer solids were reviewed by Smets [5] in 1983, our discussion below will be limited to more recent advances, putting emphasis on the explanation of non-homogeneous progress of reactions in terms of the distribution of local free volume in matrix polymers.

5.1 Spiropyran and Other Photochromic Compounds

The photoisomerization of spiropyran and other photochromic compounds or groups in polymer films has called increasing attention in connection with its potential applicability to an erasable photomemory system [2, 5]. Colorless spiropyran derivatives (SP form) isomerize to the ring-opened blue violet merocyanine form (MC) by ultraviolet light irradiation,

a $R_1 = CH_3$, $R_2 = H$, $R_3 = NO_2$
b $R_1 = Ph$, $R_2 = Br$, $R_3 = NO_2$

and the merocyanine form (MC) is discolored to the spiropyran form (SP) by heating or visible light irradiation.

Many authors since Gardlund et al. [25, 102] have devoted themselves to the thermal decoloration of the merocyanine-form of spiropyran [5, 35, 53, 62, 66, 74]. The decoloration rate in polymer films above T_g obeys first-order kinetics [102], but deviates from it below T_g and becomes much smaller than in homogeneous solvents [35, 102], as illustrated in Fig. 10. The polarity of solvent [103] and matrix also very much influences the rate of thermal decoloration of spirobenzopyran. The molecular size dependence of the decoloration rates of spiropyrans dispersed in or chemically bound to matrix polymers was also reported [62].

Kryszewski et al. [35] first proposed the idea of a kinetic matrix effect which related the non-exponential decoloration rate of the merocyanine form of spirobenzopyran (MC (a) and (b)) in PMMA to the distribution of free volume in the matrix, though it had been suggested [5, 53] that the deviation from first-order kinetics in the photo-

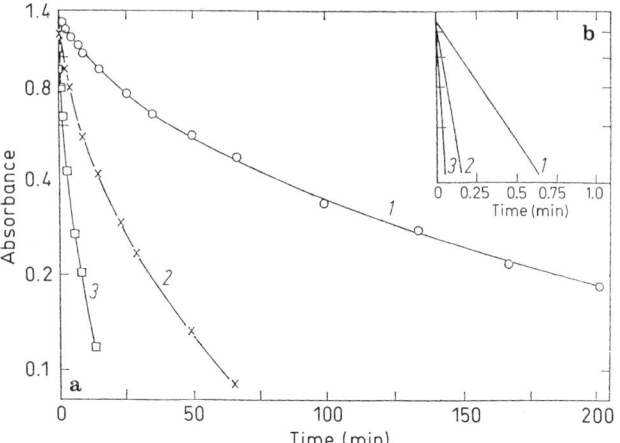

Fig. 10a and b. Kinetic plots for the ring closure reaction of merocyanine form of spirobenzopyran (MC (b)) in **a** poly(methyl methacrylate), at different temperatures: ○, 17.0; ×, 30.2 and □, 42.0 °C and **b** the curves for the reaction in methyl methacrylate at the same temperatures as those used for polymer films [35]

chromism of spiropyran arose from different metastable forms of MC, each of which is incapable of interconversion. Their proposal [35] was based on the fact that the deviation was also observed for the isomerization of azobenzene systems in which the photo-isomerized cis-form exists only as one isomeric form [36, 37]. By assuming that equilibrium exists between the expanded and collapsed forms of the microenvironment around the merocyanine (MC) and that thermal decoloration occurs only from the expanded form, $MC_{expanded}$ (Eq. (3)), they reproduced [35] the experimentally-observed

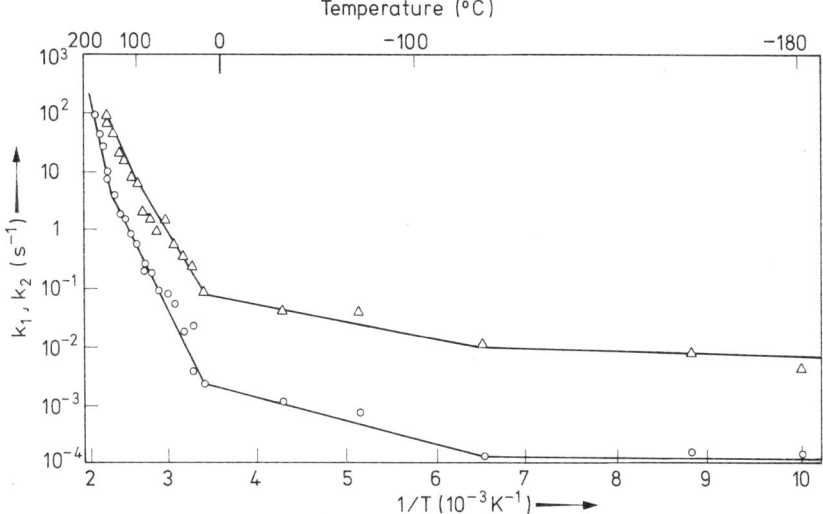

Fig. 11. Arrhenius plots for the photo-induced decoloration rate coefficients k_1 (○) and k_2 (△) in spiropyran-doped polycarbonate film [51]

$\sqrt{t}$ dependence of the decoloration reaction with an appropriate profile of density distribution function.

Horie et al.[51] measured the rate of photodecoloration by monitoring light irradiation at 560 nm of MC (a) in polycarbonate film after the coloration by a single laser pulse at 337 nm over the wide temperature range 80–453 K. The coloration of spiropyran completed within several nanoseconds. The photo-induced decoloration of the merocyanine form (MC (a)) in the film proceeded exponentially for $T > T_g$, but deviated from single exponential for $T < T_g$. Arrhenius plots of the apparent two rate coefficients for the decoloration process showed breaks at T_g, T_β and T_γ of the matrix polycarbonate (Fig. 11). The coloration of spiropyran in polycarbonate proceeded for $T < T_\gamma$ and even at 4 K [34], and the structure of the colored species for $T < T_\gamma$ was considered to have a conformation (MC') which needs no internal rotation after the bond dissociation of the spiropyran forms. Thus the existence of apparent slow and fast decolorations even for $T < T_\gamma$ cannot be attributed to the formation of two types of isomers of the merocyanine form. Figure 11 suggests that the decoloration process of spiropyran in polymer film should be related to the inhomogeneous distribution of free volume and the molecular motion of the matrix polymer.

(MC')

Richert et al.[66, 74, 75, 104] analyzed thermal decoloration of the merocyanine form of spirobenzopyran in PMMA as a dispersive process, and compared the experimental data with the stretched exponential equation (Eq. (6)) and the integral equation for uncoupled individual first-order reactions as a function of activation energy ε

$$R(t) = \int_0^\infty G(\varepsilon) \exp[k(\varepsilon) t] \, d\varepsilon$$

$$= \frac{R(0)}{\sqrt{2\pi\sigma^2}} \int_{-\infty}^\infty \exp\left[\frac{-(\varepsilon_0 - \varepsilon)^2}{2\sigma^2}\right] \exp[-k(\varepsilon) t] \, d\varepsilon \quad (25)$$

In Eq. (25) the Gaussian distribution of activation energy ε whose width σ decreased with increasing temperature was introduced for the explanation of the non-exponential dependence of the reaction profile R(t). Although thermal decoloration of the merocyanine form of spiropyran in PMMA [66] gave a good agreement to the stretched exponential equation (Eq. (6)) with $\alpha = 0.366$ if applied to a limited period of time, the concept of uncoupled parallel relaxations (Eq. (25)) also provided an appropriate description of the reaction profile. The gradual decrease in the dispersion of reaction rates with rising temperature was associated with the time for which a chromophore memorizes its initial matrix cage [104] for the decoloration of MC in several poly(alkyl acrylate) samples.

The photochromic reaction of spirobenzopyran is accompanied by a marked change in polarity owing to the ion pair formation. The spiropyran with long alkyl chains undergoes photo-induced changes in the wettability of the films [105], and also induces the formation of J-aggregate in Langmuir-Blodgett film [106] under ultraviolet irradiation with a sharp and intense J-absorption band at 619 nm. The photoinduced stacking or crystallization of spiropyrans attached to polymer side chains was also observed [107, 108], and the mechanism of photoinduced aggregation of side chain spirobenzopyran groups was studied by laser flash photolysis technique [109]. Photoirradiation induced the change in surface pressure of the monolayer of PMMA with spirobenzopyran side groups [110] as well as the change in optical rotatory power of poly(L-glutamate) thin film containing spirobenzopyran [111].

Thermal backward reactions of photochromic spiroindolinonaphthoxazine (SNOX) [73] and salicylidenaniline [112] in polymer matrices were found to be dispersive processes with an $\exp(-t^\alpha)$ dependence. The dispersion parameter α is plotted against T/T_g in

Fig. 12 [73] for thermal decoloration of spiroindolinonaphthoxazine in PMMA, cyclized isoprene rubber (CIR), and poly(2,6-dimethylphenylene oxide) (PPO). It takes values close to unity above the T_g of the matrix polymers. Around T_g, α rapidly decreases with decreasing temperature, reaching a value of about 0.5 below T_g for the samples colored by continuous xenon lamp irradiation. This value of α corresponds to

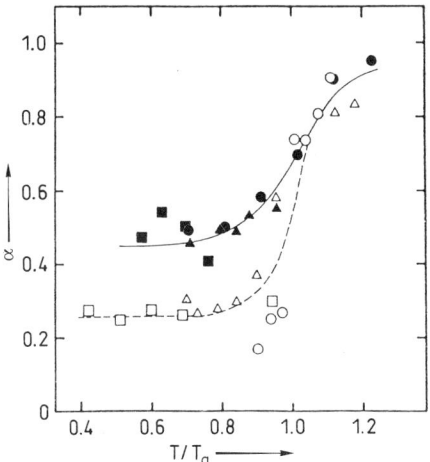

Fig. 12. Dispersive parameter α plotted against T/T_g for thermal decoloration of spironaphthoxazine (SNOX) in three polymer matrices [73]. ●, ○: CIR; ▲, △: PMMA; ■, □: PPO. Filled symbols represent continuous Xe light coloration and unfilled was laser pulse coloration

the thermal decoloration kinetics of Kryszewski et al.[35] for spiropyrans in PMMA. The unfilled symbols in Fig. 12 represent the α values when photocoloration was induced by laser pulse irradiation. Values of α clearly different from those values for samples with xenon lamp coloration were obtained below T_g. This finding suggests that the difference in photocoloration process must be carefully examined for understanding thermal reactions of photochromic species. The distribution of reaction rates is much broader in colored species produced by laser pulse than in those produced by continuous xenon light. Various conformations of colored species might not reach equilibrium immediately after laser pulse irradiation, giving smaller values of α = 0.2–0.3. On the other hand, the decoloration process after continuous xenon lamp irradiation would obey kinetics starting from the equilibrium conformational states of the colored species.

Several phenyl or heterocyclic fulgides with methyl substituents such as (FG) were reported by Heller et al.[113-115] as organic fatigue-resistant photochromic compounds.

(FG, colorless) ⇌ (FG, colored)

The photodecoloration of the cyclized form of fulgide in PMMA film proceeds following first-order kinetics, and its quantum yield (Φ = 0.06)[116] is the same as that in toluene solution[114]. This fact suggests a very small critical free volume v_{fc}, and also suggests that the ring-opening decoloration reaction is not restricted by the mobility of the matrix polymer at room temperature. The apparent rate of photodecoloration

Table 6. Quantum yields for forward $\Phi_{A \to B}$ and backward $\Phi_{B \to A}$ photochromic reactions and relaxation time τ for thermal backward reactions at room temperature

Photochromic molecule	Medium	$\Phi_{A \to B}$		$\Phi_{B \to A}$		$\tau = 1/k_{thermal}$
Spiropyran[103] (SP)	Ethanol	0.12		0.10		0.9 h
	2-Propanol	0.3 ~ 0.4		0.02 ~ 0.04		
	Toluene	0.60		0.10		
	Polycarbonate[34]	0.15				4.0 h
Azobenzene[38]		365 nm	440 nm	365 nm	440 nm	
	Ethanol	0.12	0.15	0.23	0.55	280 h
	Ethyl acetate	0.09	0.25	0.14	0.48	230 h
	Polycarbonate	0.07	0.22	0.15	0.49	100, 200 h
Fulgide (FG)	Toluene[114]		0.20	0.06		
	PMMA[116]			0.06		14 month
Dihydropyrene[117]	n-pentane	0.013				~50 h
	PMMA	~0.01				~50 h

A → B: Forward reaction from stable to metastable isomers.
B → A: Backward reaction from metastable to stable isomers.

of fulgide at 4 K appears similar to that at room temperature, and the extent of photodecoloration is limited to about 13% at 4 K [34]. Nearly equal rates or quantum yields for mobile sites at 4 K and 298 K suggest the absence of the cage effect for the ring-opening reaction of the colored form of fulgide. On the other hand, quantum yield of ring-opening, photocoloration of spirobenzopyran at 4 K was about 200 times smaller than that at room temperature in polycarbonate film owing to the dominant occurrence of cage backward reaction. The difference between these two cases [34] could be due to the electrocyclic nature of the isomerization reaction of fulgide, that is, thermal ring-closure reaction should be disrotatory, but the disrotatory cage backward reaction does not occur after the conrotatory photoinduced ring-opening reaction of fulgide.

Typical quantum yields for forward ($\Phi_{B \to A}$) and backward ($\Phi_{B \to A}$) photo reactions and relaxation time $\tau = 1/k_{thermal}$ for thermal backward reaction of photochromic molecules at room temperature are given in Table 6.

5.2 Azobenzene and its Derivatives

Many studies have been made on the photochemical trans to cis isomerization [26, 36, 38, 54, 55, 56, 63, 118] and thermal cis to trans backward isomerization [36, 37, 38, 54, 55, 118] of azobenzene chromophores bound to the backbone or attached to the side chain of polymers or molecularly dispersed in film. While photoisomerization of the azobenzene residue in a polymer chain occurs by a single rate process in dilute solution, it has been suggested that its rate in solid films may be separated into two parts, i.e. an initial process as fast as in dilute solution is followed by a slow second process. The fractional amount of the first part decreases with physical aging, but increases with temperature, plasticization, or tensile deformation [55]. This fraction has been related to the number of regions where local free volumes are greater than a critical size necessary for the photoisomerization of the azobenzene group [56].

Azobenzene molecularly dispersed in polymer film has a rather small critical free volume needed for isomerization, and hence photo- and thermal isomerizations of azobenzene at room temperature are generally regarded as being chemically controlled even in polymer matrices below T_g. However, an anomalous deviation from first-order kinetics was observed for the thermal cis → trans isomerization of azobenzene molecularly dispersed [38, 118] or bound to the polymer side chain [36, 37] in polymer film. For example, the rate of thermal cis → trans isomerization of azobenzene in polycarbonate film prepared by UV irradiation is faster than the rates in solutions, and it gradually approaches the value in solution as shown in Fig. 13. This anomalous fast process in the beginning of thermal backward isomerization was attributed by Paik and Morawetz [36] to trapping of some portion of cis isomer in a strained conformation. However, Eisenbach [37] considered it due to the restricted fluctuation of free volume. To eliminate such unstable cis azobenzene molecules in the polymer, we prepared a cis-rich sample cast into film from solution under continuous UV light irradiation. Its thermal isomerization reaction (● in Fig. 13) almost obeys first-order kinetics, supporting the idea that the existence of residual strain in the cis-azobenzene film sample prepared by photoisomerization is responsible for the anomalous deviation from first-order kinetics.

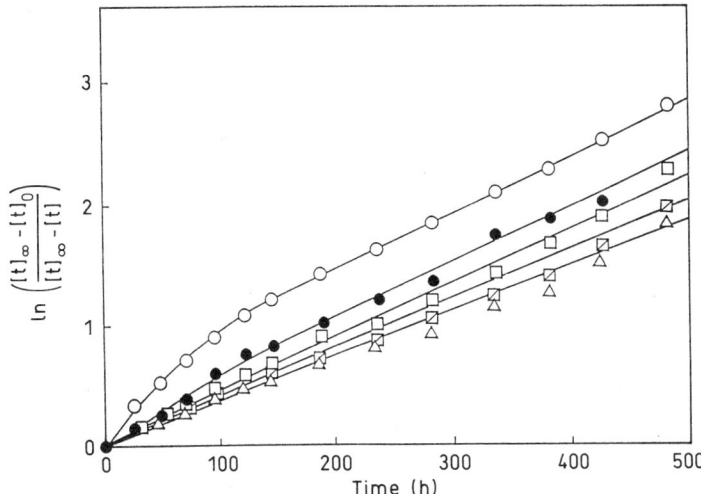

Fig. 13. Thermal backward cis → trans isomerization of azobenzene at room temperature in polycarbonate films UV irradiated in advance onto the film (○), UV irradiated during casting (●), as well as of azobenzene in ethylacetate (□), ethanol (△), and dichloroethane (⌷) UV irradiated in advance [38]

Sung et al. [56,119] measured, by repeated laser pulse irradiation, the trans to cis photoisomerization of azobenzene chromophores incorporated at three specific sites on the polystyrene chain: the chain center (C-PS), a chain end (E-PS), or a side group (S-PS). Typical changes in trans content during pulse irradiation are shown in Fig. 14. Sung et al. [56] considered that the microenvironment of the azobenzene residue

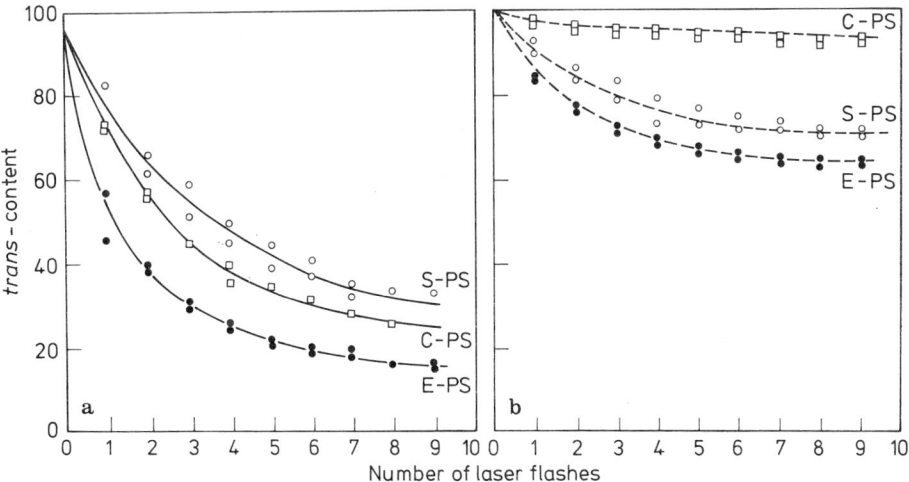

Fig. 14a and b. Trans → cis photoisomerization of azobenzene chromophores in polystyrene chains as a function of the number of laser flashes at 20 °C **a** in dilute benzene solution and **b** in the glassy state [56]

consists of freely mobile sites and frozen sites. Following Robertson [64], they related the fraction α of freely mobile sites to the area in the free volume size distribution profile where local free volumes are greater than a critical size necessary for the photoisomerization. Since the change in trans content A is expressed by Eq. (4), the reaction profiles in the glassy state were predicted by using the value of α calculated from the asymptotic trans content and that of $k + k'$ in solution. The results are shown by dashed lines in Fig. 14b. Detailed experiments with the same samples, including the effect of physical aging [119], revealed the deviation from first-order kinetics for the photoisomerization in the glassy state, and the results were analyzed on the assumption that the fast and slow processes combined are represented by

$$e^{-I'(\delta)} = \alpha e^{-A_1 t} + (1 - \alpha) e^{-A_2 t} \qquad (28)$$

Here, $I'(\delta)$ is the first-order kinetic parameter [36] proportional to the number of flashes t as a function of difference in optical densities, $\delta = D_\infty - D$, at the photostationary state (D_∞) and after t flashes (D), α the fraction of the fast process, and A_1 and A_2 the rate constants for the fast and slow processes. The values of A_1 were considered the same as the rate constant in dilute solution, and A_2 was treated as being smaller than A_1 by approximately two orders of magnitude at room temperature. No dependence of A_1 and A_2 on the labeling site to which azobenzene was attached was observed. Plots of α as a function of physical aging time at 80 °C (Fig. 15) show that the fraction of freely mobile reaction sites markedly decreases in the following order: molecularly dispersed probe > end label > side label > center label. The increase in physical aging time causes a gradual decrease in α for each sample, indicating a shift of average free volume to smaller size. The photoisomerization of azobenzene residue in epoxy resin was also used for monitoring the change in free volume distribution during the curing reaction of the epoxy resin with diaminodiphenylsulfone [120].

The temperature dependence of the rate of photochemical processes is very important since it provides information on molecular motion of the matrix polymer over a

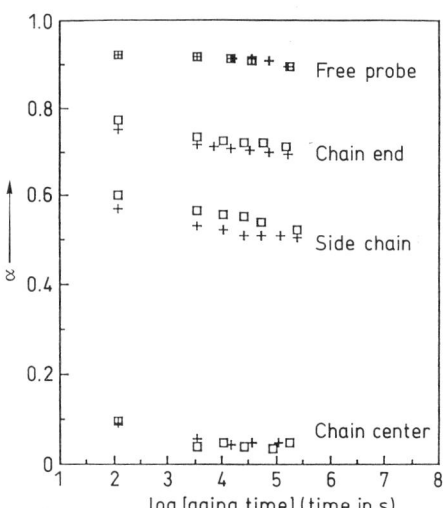

Fig. 15. Changes in the fraction α of fast photoisomerization, as a function of the physical aging time at 80 °C for three azobenzene-labelled polystyrenes and freeprobe in polystyrene [119]. (□: cooling from 120 °C to 80 °C, +: cooling from 110 °C to 80 °C)

wide temperature range. We [38] measured the trans to cis photoisomerization of azobenzene in polycarbonate film over a very wide temperature range (4 to 400 K). Figure 16 shows that the initial rate of photoisomerization does not decrease very much with decreasing temperature, but that the final fraction of trans form $x_{t\infty} = [t]_\infty/[t]_0$ increases markedly with the decrease in temperature. The temperature dependence of equilibrium or limiting trans fractions $x_{t\infty}$ is illustrated in Fig. 17. The marked increase in $x_{t\infty}$ with increasing temperature above room temperature is due to the predominant occurrence of thermal cis → trans isomerization; $x_{t\infty}$ is determined by the equilibrium of photoforward and thermal backward reactions. At lower temperatures than room temperature, there is no substantial temperature dependence of equilibrium points in ethanol solution, because there is no thermal backward reaction and only a slight photobackward reaction contributes to the equilib-

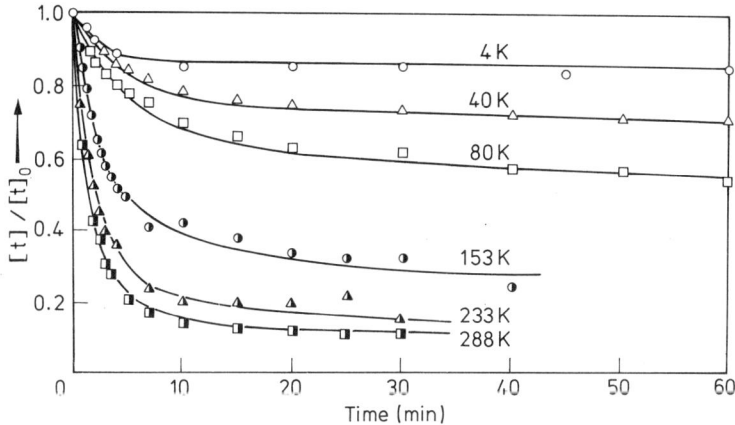

Fig. 16. Photoisomerization of trans-azobenzene in polycarbonate film at various temperatures [38]. Solid lines are from simulations according to the free volume fluctuation model

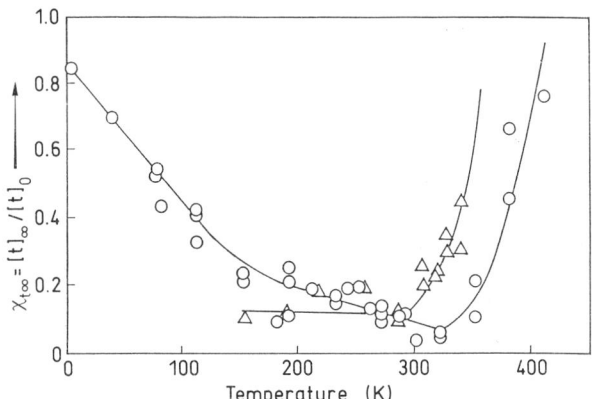

Fig. 17. Temperature dependence of final fraction $x_{t\infty}$ of trans form for photoisomerization of trans-azobenzene in polycarbonate film (○) and in ethanol (△) [38]

rium. But in polycarbonate film below room temperature, photoisomerization stops at a conversion smaller than the equilibrium value in solution, and the limiting conversion decreases with decreasing temperature. The gradual change in slope in the limiting trans fraction at about 150 K corresponds to the γ-transition of matrix polycarbonate. Small scale rotational motions of polymer phenylene groups are also thought to be restricted at temperatures below T_γ. However, even at the liquid helium temperature (4 K) trans-azobenzene photoisomerizes up to a 17% conversion. The deviation from first-order kinetics is also observed below room temperature and becomes more marked with decreasing temperature, but at 4 K the rate for trans-azobenzene at the mobile sites is almost the same as that at room temperature.

Thus, the reaction environment was divided into two sites: freely mobile sites and diffusion-controlled sites. The rate coefficients for diffusion-controlled sites do not have a single value but should be dispersive depending on the size of local free volume around azobenzene. The reaction in diffusion-controlled sites was considered to proceed with an activation energy that makes the local free volume around azobenzene equal to a certain critical free volume. The critical free volume was assumed to decrease with increasing temperature because of local fluctuations of free volume caused by local motion of the matrix polymer. The extent of reaction R(t) was given by Eq. 5. The free volume distribution G(f) and the reactivity $k(f)/k_0$ as functions of fractional local free volume f are schematically illustrated in Fig. 18. A simulation based on this kinetic model [121] showed a good agreement (solid lines in Fig. 16) with experimental data over a wide temperature range.

The analysis with this kinetic model was extended to the photoisomerization of trans-1,1'-azonaphthalene in polycarbonate film [122]. Victor and Torkelson [59] compared the photoisomerizations of azobenzene, 1,1'-azonaphthalene, stilbene, 4,4'-dinitrostilbene, 4,4'-diphenylstilbene, and thioindigo in polystyrene with those in toluene solution. The final extents of these reactions in polystyrene were related to

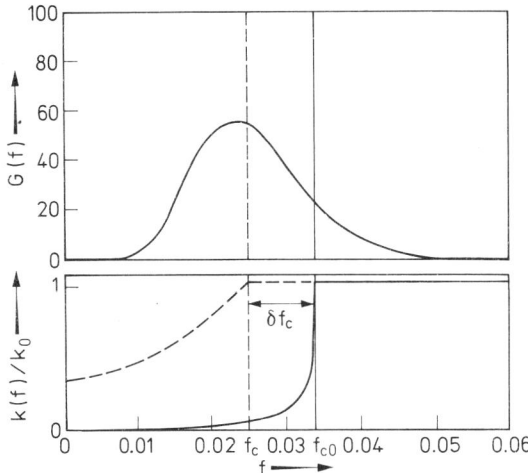

Fig. 18. Schematic illustration of fractional free volume distribution G(f) and reactivity $k(f)/k_0$ according to the free volume fluctuation model [38]. Solid line in reactivity corresponds to 4 K and dashed line to 288 K

the extra free volume needed for photoisomerization and the local free volume distribution. Thus photoisomerization of small molecules of various sizes allows us to obtain insight into the average and distribution of free volumes in amorphous polymers, which can not be evaluated unless small angle x-ray scattering (SAXS) [123, 124] or positronium annihilation spectroscopy (PAS) [125] is used.

5.3 Photochemical Hole Burning

Photochemical hole burning (PHB) is a phenomenon in which very narrow and stable photochemical holes are burned into the absorption bands of guest molecules molecularly dispersed in an amorphous solid by narrow-band excitation with a laser beam. Proton tautomerization of free-base porphyrins and phthalocyanines, and hydrogen bond rearrangement of quinizarin are typical photochemical reactions which provide PHB spectra. Recently PHB has attracted considerable interest not only as a means for frequency-domain high density optical storage, but also as a tool for high-resolution solid-state spectroscopy. Its basis as a spectroscopic tool for the study of relaxation processes in polymers and glasses [126] and its applicability to frequency domain optical storage [127] have been thoroughly reviewed. Hence, only the effect of polymer matrices on several aspects of PHB phenomena is briefly discussed below.

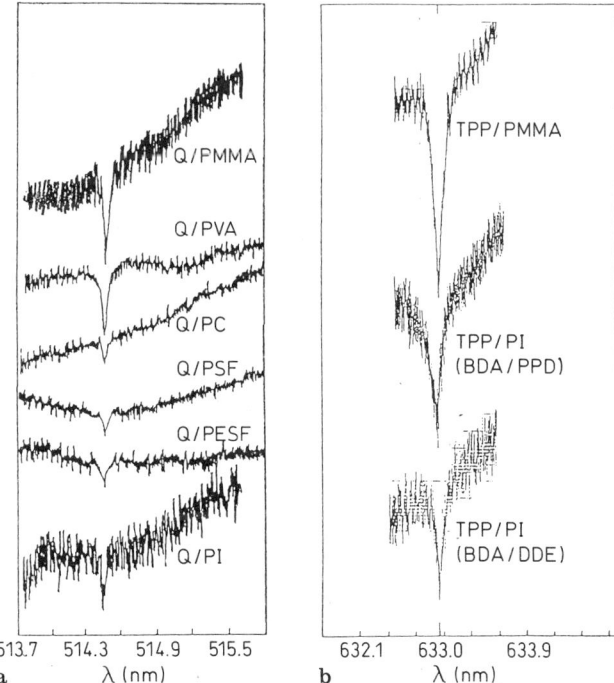

Fig. 19a and b. Photochemical hole burning (PHB) spectra of quinizarin **a** and tetraphenylporphin (TPP) **b** in various polymer matrices

Typical PHB spectra of quinizarin and tetraphenylporphin in various polymer matrices at 4 K are shown in Fig. 19. For quinizarin, narrow and stable holes are formed in PMMA and poly(vinyl alcohol) (PVA), while the hole formation is less marked in polycarbonate (PC), polysulfone (PSF), poly(ether sulfone) (PESF), and aromatic polyimide (PI) [128]. These facts suggest that the change in hydrogen bond from intramolecular to intermolecular is a key reaction for the PHB of quinizarin and

that the hydrogen-bonding ability of matrix polymers is a main factor for the efficiency of hole formation of quinizarin. For tetraphenylporphin, the efficiency of hole formation is the same for both in PMMA, PI, and in phenoxy resin (PhR) (Fig. 19b), indicating the intramolecular nature of its hole-burning phototautomeric reaction.

The hole width or homogeneous line width of $S_1 \leftarrow S_0$ transition $\Delta\omega_h$ is given by [126]

$$\Delta\omega_h = \frac{1}{2\tau_1} + \frac{1}{\tau_2} \tag{31}$$

where τ_1 is the fluorescence lifetime and τ_2 the dephasing time. The dephasing term τ_2^{-1} becomes important in amorphous matrices so that $\Delta\omega_h$ reflects the interaction between the electronically excited state and the surrounding matrix (Fig. 20). In organic crystals such as n-decane, a very narrow $\Delta\omega_h$ of the guest molecule was obtained, but it increased with increasing temperature T following a T^7 power law. The hole width $\Delta\omega_h$ seems to be proportional to $T^{1.0 \text{ to } 2.0}$ in amorphous matrices, though its absolute values are larger than those in crystals. It is noteworthy that free-base porphin in polyethylene shows a very narrow hole width despite the $T^{1.0 \text{ to } 1.3}$ dependence corresponding to the amorphous matrix [129]. This fact may be due to a non-polar structure without side groups and to the incorporation of the PHB molecule in the boundary region between paracrystallites of the polyethylene chain.

Temperature effect on electron-phonon coupling broadens an absorption line because it enhances the uncertainty of the excitation process. Spectroscopic informa-

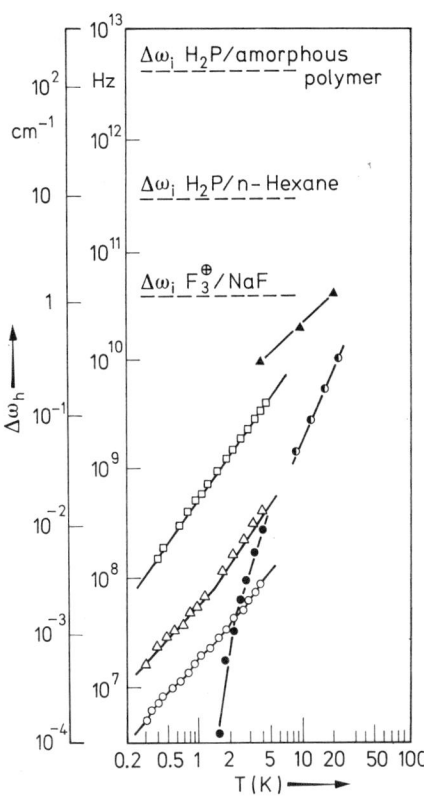

Fig. 20. Temperature dependence of (homogeneous) hole width $\Delta\omega_h$ for PHB of free-base porphin (H_2P) [129] and phthalocyanine (H_2Pc) [130]. Dashed lines correspond to (inhomogeneous) line width $\Delta\omega_i$ of absorption spectra. □: H_2P/MTHF, △: H_2P/Diglycerol, ○: H_2P/PE, ▲: H_2Pc/PMMA, ⬤: H_2Pc/PE, ●: H_2P/n-Decane

tion that has been lost owing to the broadening at elevated temperature, however, can often be recovered by recooling the system to the temperature at which the holes were originally burned. Thus, temperature-cycling experiments [131, 132] allow us to know the temperature dependence of hole stability and give information on the possibility of storing the same high information density inscribed at low temperature but archivally maintained at elevated temperature. The changes in relative hole depth, $\Delta A/\Delta A_0$, during a temperature-cycling experiment are shown in Fig. 21 for the holes of tetraphenylporphin (TPP) burned at 4 K in PMMA and in aromatic polyimide (PI) [133]. The hole formed by 16 min irradiation by a He—Ne laser (3.8 mW/cm²) at 4 K dissappears at the 30 min-annealed temperature if it is higher than 50 K, but the hole partly recovers when the sample was cooled again to 4 K. The hole depth measured at annealed temperatures disappears at 50 K for both PMMA and PI matrices, but the restored hole depth measured at 4 K for TPP in PI film is larger than that for TPP in PMMA film. This finding may be explained by the stiff structure of aromatic polyimide prepared from biphenyltetracarboxylic anhydride (BDA) and p-phenylenediamine (PPD). Figure 21 shows that partial hole recovery occurs at 4 K even after cyclic annealings at 110 K for TPP in PMMA and at 130 K for TPP in PI film.

Recently, two-color, photon-gated PHB was proposed and observed in some inorganic [134] and organic [134—137] systems. This proposal was to avoid the hole destruc-

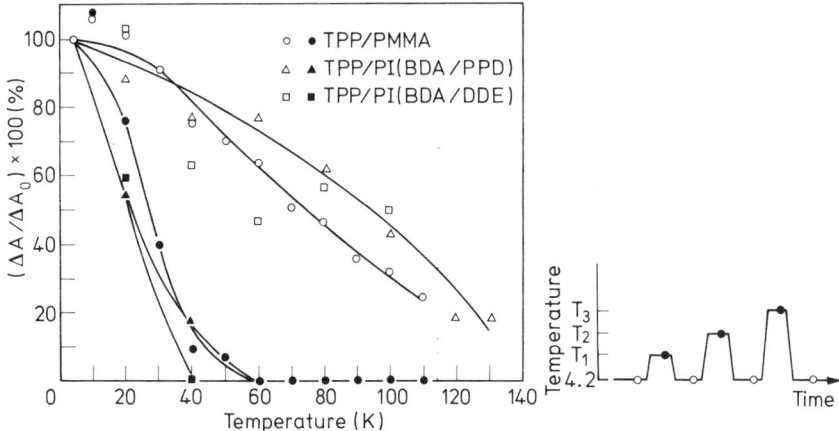

Fig. 21. Change in relative hole depth $\Delta A/\Delta A_0$ during PHB temperature-cycling experiments for TPP in PMMA and TPP in PI [133]

tion during repeated reading of hole spectra, which is inherent in usual one-photon PHB systems [138]. The mechanism assumed for the two-color excitation PHB of the carbazole molecule in boric acid glass [135] is illustrated in Fig. 22. After excitation in the singlet-singlet origin with $\lambda_1 = 335$ nm, the molecule undergoes an intersystem crossing with a high yield to form triplets. From the excited triplet state (level 3), the molecules return to the original ground state if λ_2 is not present, and no hole is formed. However, in the presence of $\lambda_2 = 360$ to 450 nm, holes are formed at the singlet excitation wavelength λ_1 presumably owing to photoionization of the molecule in the triplet manifold and also to trapping of the ejected electron in the boric acid glass matrix. Other two-color PHB systems observed are Sm^{2+} in BaClF [134], Sm^{2+} in CaF_2 [139], photoadducts of anthracene and tetracene in PMMA [136], and a donor-acceptor complex of zinc tetrabenzoporphyrin with chloroform in PMMA [137].

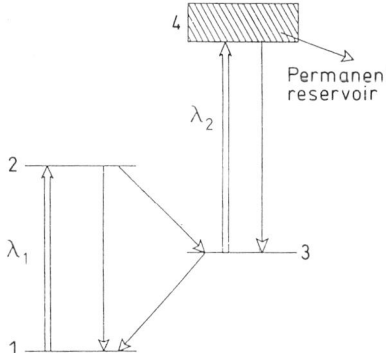

Fig. 22. Illustration of two-color, photon-gated photochemical hole buring (PHB) for a four-level system [135]

6 Photodimerization

It has been known for long that some unsaturated molecules such as trans-cinnamic acid dimerize through a photocycloaddition reaction in the crystalline state [21, 140]. Polymorphic modifications of a given compound show significant differences in chemical behavior, and the controlling factor for the reactions in the crystalline state has been found to be topochemical requisites, which are relatively fixed distances and orientations between potentially reactive groups.

Incorporation of two photodimerization groups in a monomer molecule gives the possibility of preparing a linear polymer, and the binding of photodimerizable groups to the side chain of the polymer produces a crosslinked polymer which is practically important as a negative-type photoresist.

Many bifunctional monomers have been reported to polymerize and are summarized in some review articles [9, 141, 142]. Especially, intensive studies have been made on the polymerization mechanism and the mechanism and crystallography of 2,5-distyrylpyrazine and other diolefin crystals [9]. Photopolymerization of diacetylenes in crystals and in Langmuir-Blodgett-multilayers have attracted recent interest, because the products are totally conjugated systems and have unusual optical, electrical, and reaction properties [10, 142–144]. The atomic positions in the monomer are compared with those in the polymer crystal in Fig. 23 for 2,5-distyrylpyrazine [9] and in Fig. 24 for 1,6-di-(N-carbazolyl)-2,4-hexadiyne [144]. In the polymerization of the former, the same crystal structure is maintained before and after photopolymerization, and a small displacement of a monomer unit in the direction of chain growth results in a contraction of 1.8% in the direction of C-axis together a crystal shrinkage of 1%. The reaction distance between neighboring double bonds is 3.7 Å in this case [9]. The lattice parameters for the crystal structure changed during the polymerization of diacetylene derivatives [144].

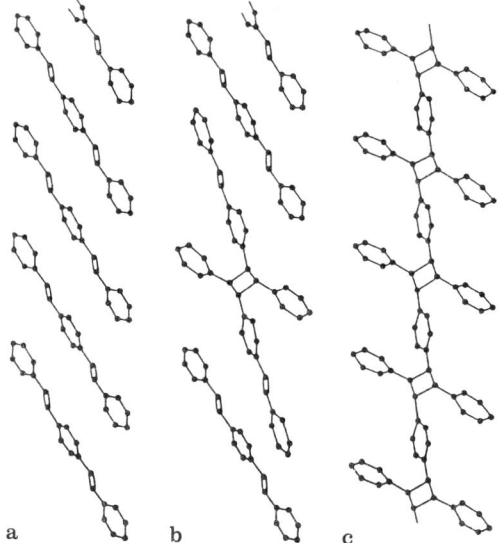

Fig. 23a–c. Schematic illustration of the conversion of **a** 2,5-distyrylpyrazine into **b** dimer and **c** polymer [9]

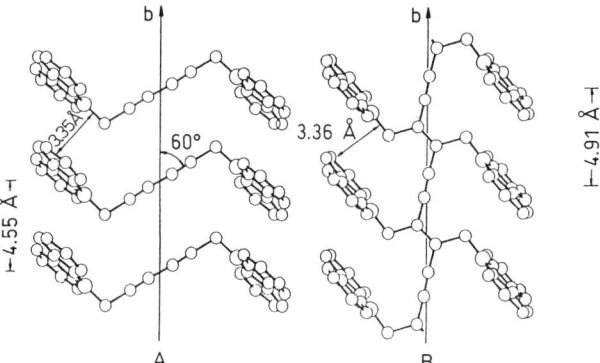

Fig. 24. Projection of the dicarbazolylhexadiyne monomer and polymer crystal structures on the plane of the polymer chain [144]

Poly(vinyl cinnamate) is the earliest synthetic photo-polymer [145]. Its molecular structure, the polyvinyl backbone with cinnamoyl side chains, was proposed with the expectation that the photocycloaddition of polymer-bound cinnamoyl groups would crosslink adjacent macromolecules. The practical aims of the inventors were realized, and the whole range of successful photopolymers has been developed from the original idea [146]. However, the actual mechanism of crosslink formation in these systems remained unclarified until recently. Repeated attempts to identify cyclobutane derivatives in irradiated films of poly(vinyl cinnamate) failed with traces of α-truxillic acid detected in the hydrolyzed material only at the very early stage of irradiation. Schmidt [21] questioned the possibility of cycloaddition in amorphous polymer matrices on the basis of the stringent steric conditions required for the cyclization process. In fact, several observations seemed to support this view. For example, ESR signals of well-defined radicals were detected [147] in irradiated films of poly(vinyl cinnamate), suggesting a radical-based mechanism for the crosslinking process.

Egerton et al. [29] tried to obtain direct evidence for or against the occurrence of photocycloaddition by hydrolyzing a crosslinked film of poly(vinyl cinnamate) and searching for cyclobutane derivatives in the hydrolysate. It was found that at least 65% of the photoproducts were cyclodimers. The remaining part was oligomers which could not be fully characterized. This finding is not surprising because, as already discussed in Chapt. 3, the local mode molecular motion of the main chain and the rotation of the side chain of poly(vinyl cinnamate) are not frozen-in at temperatures below T_g but above T_β and T_γ.

Solid matrices affect the efficiency of photodimerization as illustrated in Fig. 25 [29]. A decrease in quantum yield for photodimerization in a solid film of poly(vinyl cinnamate) was observed, along with a virtual standstill of reaction below 50% conversion of the cinnamate group. This is in contrast to a constant quantum yield for the reaction in the crystal of cinnamic acid (A). Since all chromophores in the system have the same structure, the lack of reactivity in some of the chromophores must be due to their environment. Thus, the reactive site distribution profiles in amorphous solids were represented by using a histogram technique [29, 30]. Egerton et al. [29] considered that the reactive site distribution is static (which means that each site does not change from

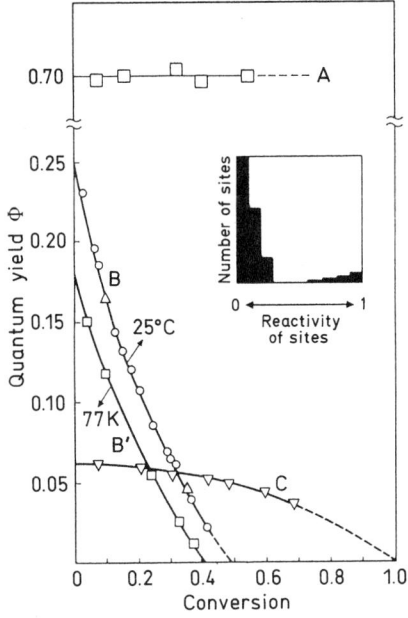

Fig. 25. Change in quantum yield φ during photodimerization of poly(vinyl cinnamate) (PVCm). (A) Crystalline cinnamic acid, (B) PVCm film, (C) PVCm solution in dichloroethane [29]. Insert: Histogram for site reactivity distribution corresponding to Curve B

nonreactive to reactive and vice versa during the reaction) because of the aggregation of polymer chains, but the formation of crosslinks would also change the reactivity during the reaction.

The photodimerization of anthryl groups bound to various polyesters or polyesterurethanes as side groups was studied by Tazuke et al. [43], who showed that segment mobility is more important for solid state reactions than the local concentration of anthryl groups. The photodimerization rate increased markedly above T_g. Polyesters or polyesterurethanes having dianthracene units in the main chain were photolyzed with $\lambda \leq 300$ nm to anthoate groups (Fig. 26) [44], and their structures were estimated from absorption and fluorescence spectra. The mobility of terminal segments depended on the polymer structure. Polyesterurethanes well retained the sand-

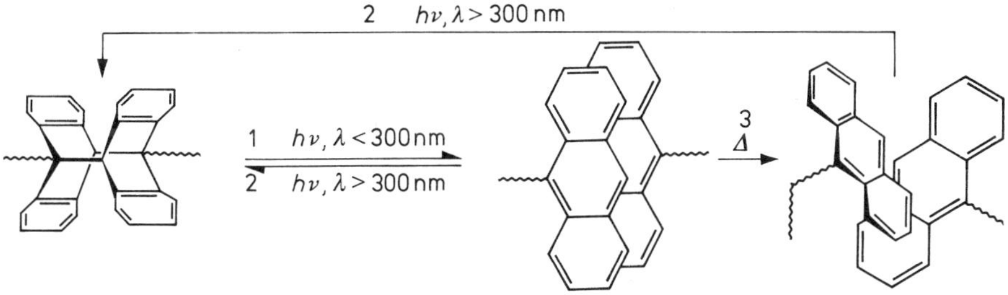

Fig. 26. Schematic illustration of photolysis of dianthracene units in polymer main chain and subsequent reactions [44]. 1: photolysis of dianthracene main chain, 2: remaking of dianthracence units, 3: thermal destruction of sandwich dimer structure

wich dimer structure at room temperature, whereas the polyester did not maintain the paired structure. The restricted mobility of terminal groups was made apparent in the remaking of dianthracene units caused by irradiating the anthroate groups at 365 nm. This remaking in polyesterurethane at 20 °C with 365 nm after photolysis at 20 °C with 300 nm occurred very rapidly, but the annealing of the photolyzed sample at 80 °C for 2 h substantially inhibited the remaking of dianthracene units at 20 °C. The effects of the history of sample preparation and the annealing on reactivity manifest themselves much more markedly in the intermolecular dimerization than in the intramolecular isomerization discussed in Chapt. 5.

The photodimerization of pendant cinnamoyl groups in cyclized polydienes is also affected greatly by the mobility of the cinnamate groups [45]. A linear relation was found between the logarithm of the photodimerization rate and $1/(T - T_g + 50)$ above T_g.

7 Chain Scission and Crosslinking Reactions

7.1 Norrish Type I and Type II Reactions

The study of photochemistry of polymeric ketones made by Guillet et al. [1, 24, 33] gives an interesting illustration of the effect of molecular mobility on two types of photochemical reactions. One is the Norrish type I process which involves a separation of radical pairs (Eq. (32)). The other is the Norrish type II process which involves intramolecular hydrogen abstraction by a cyclic six-membered intermediate, yielding a biradical which subsequently produces a ketone and an olefin (Eq. (33)).

$$-\overset{O}{\underset{\parallel}{C}}-CH_2- \underset{k_d}{\overset{h\nu}{\rightleftarrows}} -\overset{O^*}{\underset{\parallel}{C}}-CH_2- \overset{k_{CS}}{\longrightarrow} -\overset{O}{\underset{\parallel}{C}}\cdot + \cdot CH_2- \qquad (32)$$

$$(33)$$

For ethylene-carbon monoxide copolymers, the formation of the free radical type I products is suppressed to only about 10% of the total reactions because both radicals are polymeric and hardly ever escape from the immediate environment [24]. On the other hand, for ethylene-methyl vinyl ketone copolymers, photolysis yields one polymeric and one small acetyl radical, with the efficiency of type I reaction eight times as much as that for ethylene-carbon monoxide copolymers. This is because the same amount of free volume will permit substantial diffusion of the small acetyl radical away from its larger polymeric counterpart [148].

The quantum yield of the Norrish type II process in the ethylene-carbon monoxide (1%) copolymer at room temperature (Φ_{II} = 0.025) and the lifetime of the triplet excited state are comparable to those in hydrocarbon solution under the same conditions [24]. This finding can be explained by the fact that at room temperature the copolymer is above its glass transition temperature and hence a large scale movement of chain segments is possible even though the center of mass of the polymer remains fixed. When the polymer is cooled, the quantum yield begins to drop below $-40\ °C$ and reaches zero at $-120\ °C$. Since T_g of polyethylene is reported to be -80 to $-90\ °C$ [149], no occurrence of the Norrish type II reaction in the polyethylene chain should be associated with the freezing-in of local mode relaxation of the main chain.

The Norrish type II reaction of vinyl ketone copolymers also leads to scission of the backbone of polymer chain. Typical temperature dependence of the quantum yield for chain scission due to the Norrish type II reaction is shown in Fig. 27 for the copolymer of styrene and phenyl vinyl ketone [33]. There is a relatively small increase in the quantum yield below T_g owing to the presence of local mode relaxation in the main chain, but a drastic increase takes place in the region of glass transition temperature. Above T_g the quantum yield for the type II process is identical to that observed in solution.

The rate coefficient k_1 for intramolecular hydrogen abstraction in Eq. (33) is expressed by

$$k_1 = \frac{4\pi RD[H]}{1 + 4\pi RD/k_c} \tag{34}$$

which includes the both diffusion-controlled case ($k_1 \simeq 4\pi RD[H]$) and the chemically-controlled case ($k_1 = k_c[H]$). Here R is the reaction radius, D the local diffusion coefficient of the reacting carbonyl triplet and the hydrogen atom, and k_c the inherent

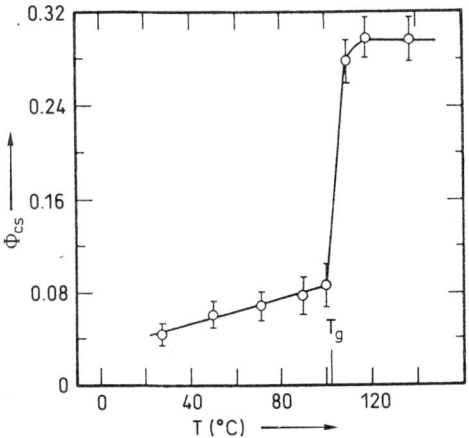

Fig. 27. Quantum yield of Norrish type II reaction of styrene-phenyl vinyl ketone copolymer film as a function of temperature [33]

bimolecular rate coefficient for hydrogen abstraction. The quantum yield Φ_{cs} for the Norrish type II chain scission is given by

$$\Phi_{cs} = \frac{k_1}{k_1 + k_d} \cdot \frac{k_2}{k_2 + k_3} \tag{35}$$

where k_d, k_2 and k_3 are the rate constants for deactivation of the excited date, β-scission of intermediate biradical, and backward hydrogen transfer, respectively. The temperature dependence of k_1 and Φ_{cs} is schematically shown in Fig. 28, where $k_1 = 4\pi RD[H]$ holds in the low temperature region and $k_1 = k_c[H]$ in the high temperature region. The temperature dependence of Φ_{cs} in Fig. 28 well describes the experimental results if it is assumed that the crossover from the diffusion-controlled to chemically controlled mechanism occurs just above T_g.

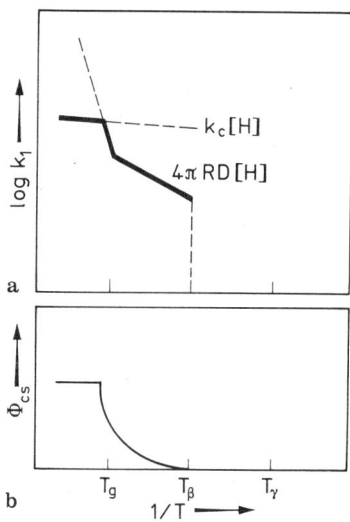

Fig. 28a and b. Schematic illustration of temperature dependence of k_1 **a** and Φ_{cs} **b** for Norrish type II reactions in polymer solids

7.2 Degradation of Vinyl and Aromatic Polymers

Many reviews and books [150-154] on the degradation of vinyl polymers have been written from kinetic, mechanistic, and practical points of view. The present section is therefore limited to a discussion on the temperature dependence of chain scission and crosslinking reactions as well as the ratio of their rates in polymer solids subject to photo- or γ-ray irradiation.

In contrast to the thermal degradation of vinyl polymers at high temperature where chain scission dominates owing to repetitive hydrogen abstraction and β-scission of polymer radicals [155], photodegradation of many solid polymers including polystyrene, poly(methyl acrylate), poly(ethyl acrylate), poly(vinyl acetate), and polyacrylonitrile at ambient temperatures yields the formation of crosslinking in the absence of oxygen [153]. Only in such polymers as poly-α-methylstyrene, poly(methyl methacrylate), and poly(phenyl vinyl ketone) does the main-chain scission and the decrease in molecular weight become dominant.

The quantum yield Φ_s, for chain scission and that Φ_x for crosslinking can be calculated from a sol-gel analysis using the Charlesby-Pinner equation [156)]

$$S + \sqrt{S} = Z_s/Z_x + 1/(Z_x t) \tag{36}$$

where S is the soluble fraction after time t, and Z_s or Z_x the rate of scission or crosslinking per initial macromolecule. Z_s and Z_x can be transformed to Φ_s and Φ_x, respectively, by knowing the incident light intensity. The values of Φ_s and Φ_x are also calculable from the change in molecular weight distribution, using the equations

$$\overline{M}_n = \overline{M}_{n,0}/(1 + Z_s - Z_x) \tag{37}$$

$$\overline{M}_w = \frac{2\overline{M}_{n,0}[Z_s - 1 + (1 + Z_z/\sigma)^{-\sigma}]}{Z_s^2 - 4Z_x[Z_s - 1 + (1 + Z_s/\sigma)^{-\sigma}]} \tag{38}$$

where $\overline{M}_n$ and $\overline{M}_w$ are the number and weight average molecular weights, respectively, $\overline{M}_{n,0}$ the initial value of $\overline{M}_n$, and σ a parameter characterizing the width of the initial molecular weight distribution of the polymer:

$$\sigma = \overline{M}_{n,0}/(\overline{M}_{w,0} - \overline{M}_{n,0}) \tag{39}$$

For the most probable distribution Eq. (38) gives

$$\overline{M}_w = \frac{\overline{M}_{w,0}}{1 + Z_s - 4Z_x} \tag{40}$$

which shows that the change in molecular weight during degradation accompanying simultaneous chain scission and crosslinking depends on whether $Z_s/Z_x = \Phi_s/\Phi_x$ is equal to, greater than or smaller than 4.

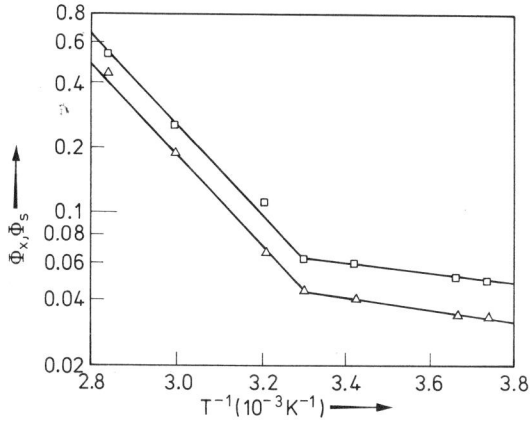

Fig. 29. Arrhenius plot for quantum yield for photodegradation of poly (vinyl acetate) as a function of temperature [158)]. △: crosslinking, □: chain scission

Typical examples of Φ_s and Φ_x including their temperature dependence are shown in Fig. 29 for photodegradation of poly(vinyl acetate) (PVAc) irradiated by 154 nm light [158]. A similar temperature dependence was observed for both Φ_s and Φ_x with $\Phi_s/\Phi_x = 1.2$ to 1.5, which indicates the predominant occurrence of crosslinking. The break at 30 °C in Fig. 29 corresponds to T_g of PVAc. It has been proposed that crosslinking results from intermolecular tertiary hydrogen abstraction by an excited acetate group followed by recombination of the geminate radicals produced. However, there is another proposal that photochemical chain scission in PVAc is due to intramolecular tertiary hydrogen abstraction through a seven-membered ring transition state [158]. We note that the intermolecular chain transfer processes may also result chain scission of PVAc for $T > T_g$, considering the high Φ_s value in Fig. 29.

The temperature dependence of the G-values G_s and G_x for scission and crosslinking of polystyrene under γ irradiation is given in Table 7 [159]. The ratio G_s/G_x increases from 0.02 at 30 °C to 2.8 at 150 °C. This increase is due mainly to a tenfold increase in G_s with G_x decreasing slightly. These results are compatible with the increased disproportion of chain scission radicals relative to their combination, and analogous to the temperature dependence of mutual termination in free radical polymerizations.

The Arrhenius plot of the G-value for radiation induced degradation of polyisobutene [160] showed breaks at T_g (-73 °C) and a secondary transition (-190 °C) with larger slopes at higher temperatures. In the region above T_g, a bimolecular chain transfer (hydrogen abstraction) reaction may occur between primary radicals formed by C—H bond rupture and secondary terminal radicals and may result in main-chain β scission. Since the inherent rates of these processes are very fast, the overall reaction rate is diffusion controlled, and depends on the mobility of polymer segments, which is not large enough for the equilibrium reactant concentration to be attained even for $T > T_g$. The proportion of cage recombination of primary radicals also becomes small at higher temperatures. Thus, the G-value for chain scission increases with increasing temperature above T_g. The crossover from a diffusion-controlled to an activation-controlled mechanism in the hydrogen abstraction from poly(vinyl alcohol) by the benzophenone triplet was evidenced by a convex break in Arrhenius plots at 110 °C [48]. Poly(methyl methacrylate) films undergo chain scission on UV-irradiation at ambient temperatures, while poly(methyl acrylate) films become rapidly insoluble owing to crosslinking. For the copolymer of methyl methacrylate (MMA) and methyl acrylate (MA) Z_x decreased with increasing MMA content, but Z_s showed a concave curve with a minimum at 33% of MMA content. The latter is due to the

Table 7. Temperature dependence of scission and crosslinking yields for γ irradiation of polystyrene of narrow molecular weight distribution irradiated in vacuum [159]

Temperature (°C)	G_s	G_x	G_s/G_x
30	0.0086	0.043	0.02
100	0.044	0.036	1.45
150	0.074	0.027	2.75

balance between the increase in T_g with increasing MMA content and the inherent higher effeciency of main-chain scission for PMMA component [161].

Most of the main-chain aromatic polymers have their absorption bands at wavelengths λ longer than 300 nm, and usually they have some functional groups which are photochemically reactive. Consequently, they are not very stable photochemically, though very noted for thermal stability. The extent of photodegradation at a certain wave length is equal to the quantum yield of photoreaction multiplied by the molar extinction coefficient. Hence, the fact that photodegradability of polysulfones [162] is greater than that of polycarbonate may be attributed to a larger molar extinction coefficient of polysulfone at $\lambda > 300$ nm, though its quantum yield for photoreaction is small. In contrast to the thermal degradation, the photodegradation of aromatic polymers at room temperature usually causes main-chain scission [147] and reduces the molecular weight. Photorearrangements such as photo-Fries and photo-Claisen reactions cause coloration. The effect of oxygen is not as marked for aromatic polymers as for photodegradation of olefins and vinyl polymers.

The quantum yields for the photo-Fries rearrangement of various aromatic polyamides in the solid state [163] are reported to be $\Phi_r = 2 \times 10^{-7} - 1 \times 10^{-6}$ (Table 8), which are hundred to thousand times smaller than Φ_r in solution, and Φ_r is smaller for polyamides with higher glass transition temperatures. Thus the mobility of polymer chains seems to be a determining factor for Φ_r of photo-Fries rearrangements.

The photodegradation of polyethersulfone proceeds mainly through main-chain scission at room temperature, but the crosslinking becomes appreciable with increasing temperature and overwhelming above 170 °C, as shown in Fig. 30 [164]. The value of Φ_s in solid film is 40 times smaller than Φ_s in solution, indicating a marked cage effect on the homolytic bond dissociation in polymer solids below T_g. The temperature at the break in Fig. 30 is about 55° lower than T_g of polyethersulfone. Some change in the mode of local molecular motion below T_g should be responsible for the change in activation energy. The quantum yield Φ_{SO} for SO_2 evolution also shows a break at the same temperature.

For polymers containing crystalline and amorphous parts, the rate of degradation decreases with increasing crystallinity when the degradation occurs mainly in the amorphous part as is the oxidation of polyethylene [165]. In contrast to this, oxidative degradation of polyoxymethylene and polypropylene under γ-irradiation is not much

Table 8. Quantum yields of photo-Fries rearrangement Φ_{Fries} of aromatic polyamides [163]

Aramid	Abbreviation	T_g °C	Initial Φ_{Fries} mole/einstein × 10^7
Poly(m-benzamide)	PmBA	265	9.6 ± 3
Poly(m-phenylene isophthalamide)	PmPiPA	275	10.7 ± 3
Poly(p-phenylene isophthalamide)	PpPiPA	306	6.3 ± 2
Poly(m-phenylene terephthalamide)	PmPtPA	334	5.1 ± 2
Poly(p-benzamide)	PpBA	>330	~1.6 ± 0.4
Poly(p-phenylene tarephthalamide)	PpPtPA	>330	ND.
PmPiPA in DMSO solution			6,000

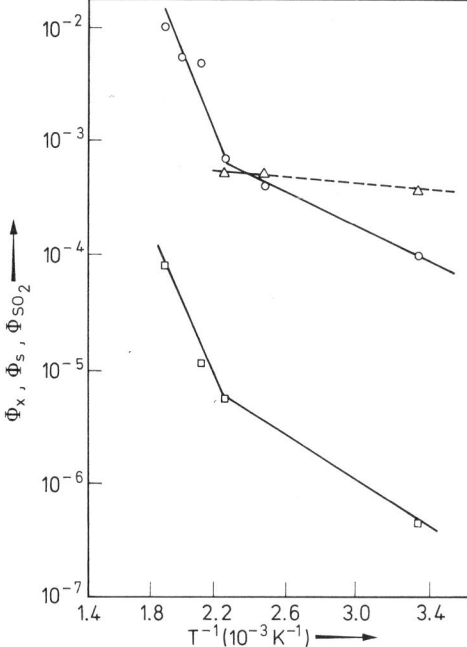

Fig. 30. Arrhenius plots for quantum yield of crosslinking, Φ_x (○), chain scission Φ_s (△), and SO_2 evolution, Φ_{SO_2} (□), for photodegradation of polyethersulfone in nitrogen atmosphere [164]

influenced by crystallinity [166], because high concentration oxidation proceeds at the crystalline-amorphous interface after it is initiated by radicals produced inside the crystalline part and transferred to the interface [167]. Bimolecular reactions of peroxyradicals in polyoxymethylene and polypropylene are not terminating-type as in polyethylene. This would explain the high concentration of radicals at the interface. Reaction kinetics which takes such a reactivity at the interface in accout has been proposed [168].

It is well known that nitric acid only reacts with the folding part of polymer single crystals. Halogenation also occurs only at the folding part [169] owing to the inhibited diffusivity of the reagents inside the crystal. The radiation crosslinking was shown to be inhibited inside the crystal [170]. The free radical can be formed inside the crystal but the conformational change accompanying crosslinking would be impossible there, and the crosslinking occurs only at the folding surface after a sequence of radical transfer reactions. Theory of usual random crosslinking can not be applied to this case [171], since the reaction proceeds only at the specific sites. The formation of vinylene which is possible by small motion is realized inside the crystalline part [172]. Photoinduced chain scission of long chain alkanone does not occur inside the crystal [173]. The radical transfer in the polyethylene single crystal was considered as a random diffusion process, and the radical disappearance obeyed Eq. (15) with the rate coefficient including the time-dependent transient term [42, 174]. It was considered that radical transfer in the polyethylene single crystal proceeds not intramolecularly along the polymer chain but intermolecularly across the polymer chain on the basis of a study with the urea adduct one-dimensional crystal [174].

8 Acknowledgement

The authors express their gratitude to Prof. H. Fujita who offered them the opportunity of writing this review article and edited the manuscript with great care.

9 References

1. Guillet JE (1977) Pure Appl Chem 49: 249
2. Williams JLR, Daly RC (1977) Prog Polym Sci 5: 61
3. Horie K, Mita I (1985) Kobunshi 34: 448
4. Mita I, Horie K (1987) J. Macromol Sci-Rev Macromol Chem Phys C27:91
5. Smet G (1983) Adv Polym Sci 50: 17
6. Winnik MA (ed) (1986) Photophysical and Photochemical Tools in Polymer Science, NATO ASI series Vol 182, Dordrecht: Reidel
7. Yamaoka H, Tagawa S (1984) Nihon Genshiryoku Gakkaishi 26: 739
8. Dilling WL (1983) Chem Rev 83: 1
9. Hasegawa M (1983) Chem Rev 83: 507
10. Sandman DL (ed) (1987) Crystallographically Ordered Polymers, ACS Symp. Series 337 Washington
11. Tabata Y, Suzuki T, Oshima K (1966) J. Macromol Chem 1: 773
12. Barkalov IM, Goldanskii VI, Enikolopyan NS, Terekhova SF, Trofimova GM (1964) J Polym Sci, Part C, 4: 909
13. Nakatsuka K, Adachi K, Suga H, Seki S (1968) J Polym Sci, Part B, 6: 779
14. Seki S (1970) Kobunshi 19: 960
15. Kaestu I, Nakase Y, Hayashi K (1969) J Macromol Sci-Chem A3: 1525
16. Kelley FN, Bueche F (1961) J Polym Sci 50: 549
17. Horie K, Mita I, Kambe H (1968) J Polym Sci Part A-1, 6: 2663
18. Ito K (1984) Polym J 16: 761
19. Mita I, Okuyama H (1971) J Polym Sci Part A-1, 9: 3437
20. Horie K, Hiura H, Sawada M, Mita I, Kambe H (1970) J Polym Sci Part A-1, 8: 1357
21. Schmidt GM (1971) Pure Appl Chem 27: 647
22. Numata S, Fujisaki K, Kinjo N (1984) Polyimides, Vol 1, Mittal KL (ed), Plenum, N.Y., p 259
23. Polym. Soc. Japan (1978) Fundamentals of Polymer Science, Tokyo Kagaku Dozin, Tokyo, p 215
24. Hartley GH, Guillet JE (1968) Macromolecules 1: 165
25. Gardlund ZG (1968) J Polym Sci, Part B, 6: 57
26. Kamogawa H, Kato M, Sugiyama H (1968) J Polym Sci part A-1, 6: 2967
27. Cowell GW, Pitts Jr JN (1968) J Amer Chem Soc 90: 2
28. Fakirov S, Schultz J (eds) (1986) Solid State Behavior of Linear Polyesters and Polyamides, Chap. 1
29. Egerton PL, Pitts E, Reiser A (1981) Macromolecules 14: 95
30. Pitts E, Reiser A (1983) J Am Chem Soc 105: 5590
31. Shibasaki Y, Fukuda K (1979) J Polym Sci, Polym Chem 17: 2947
32. Stickler M (1983) Makromol Chem 184: 2563
33. Dan E, Guillet JE (1973) Macromolecules 6: 230
34. Horie K, Hirao K, Kenmochi N, Mita I (1988) Makromol Chem Rapid Commun 9: 267
35. Kryszewski M, Nadolski B, North AM, Pethrick RA (1980) J Chem Soc, Faraday Trans II 76: 351
36. Paik CS, Morawetz H (1972) Macromolecules 5: 171
37. Eisenbach CD (1980) Ber Bunsenges Phys Chem 84: 680
38. Hirao K, Horie K, Mita I (1985) Polym Prepr Jpn 34: 1573 Macromolecules, to be published
39. Shimada S, Hori Y, Kashiwabara H (1981) Polymer 22: 1377
40. Horie K (1984) Kino Zairyo 4(10): 8
41. Wundrich K (1974) Eur Polym J 10: 341
42. Geuskens G, Borsu M, David C (1972) Eur Polym J 8: 1347

43. Tazuke S, Hayashi N (1978) J Polym Sci Polym Chem Ed, 16: 2729
44. Tazuke S, Tanabe T (1979) Macromolecules 12: 853
45. Azuma C, Sanui K, Ogata N (1980) J Appl Polym Sci 25: 1273, (1982) J Appl Polym Sci 27: 2065
46. Horie K, Morishita K, Mita I (1984) Macromolecules 17: 1746
47. Horie K, Tsukamoto M, Morishita K, Mita I (1985) Polym J 17: 517
48. Horie K, Ando H, Mita I (1987) Macromolecules 20: 54
49. Grassie N, Scotney A, Davis TI (1975) Makromol Chem 176: 963
50. Li SKL, Guillet JE (1977) Macromolecules 10: 840
51. Horie K, Tsukamoto M, Mita I (1985) Eur Polym J 21: 805
52. Mita I, Horie K (1983) Degradation and Stabilization of Polymers (Jellinek HHG (ed)), Elsevier, Amsterdam, Chap. 5
53. Miura M, Hayashi T, Akutsu F, Nagakubo K (1978) Polymer 19: 348
54. Sung CSP, Lamarre L, Chung KH (1981) Macromolecules 14: 1839
55. Lamarre L, Sung CSP (1983) Macromolecules 16: 1729
56. Sung CSP, Gould IR, Turro NJ (1984) Macromolecules 17: 1447
57. Tsapovetskii MI, Laius LA, Bessonov MI, Koton MM (1981) Dokl AN SSSR 256: 912
58. Itagaki H, Horie K, Mita I, Washio M, Tagawa S, Tabata Y, Sato H, Tanaka Y (1985) Polym Prepr Jpn 34: 1569
59. Victor JG, Torkelson JM (1987) Macromolecules 20: 2241
60. Ushiki H, Hirayanagi K, Sindo Y, Horie K, Mita I (1983) Polym J 15: 811
61. Horie K, Hirao K, Mita I, Takubo Y, Okamoto T, Washio M, Tagawa S, Tabata Y (1985) Chem Phys Lett 119: 499
62. Smet G, Thoen J, Aert A (1975) J Polym Sci Polym Symp 51: 119
63. Paik-Sung CS, Lamarre L, Tse MK (1979) Macromolecules 12: 666
64. Robertson RE (1978) J. Polym Sci Polym Symp 63: 173
65. Kohlrausch R (1947) Ann Phys (Leipzig) 12: 393
66. Richert R, Elschner A, Bassler H (1986) Z Phys Chem N F, 149: 63
67. Palmer RG, Stein DL, Abraham E, Anderson PW (1984) Phys Rev Lett 53: 958
68. Glarum SH (1960) J Chem Phys 33: 639
69. Shlesinger MF, Montroll EW (1984) Proc Natl Acad Sci USA 81: 1280
70. Bendler JT, Shlesinger MF (1985) Macromol 18: 591
71. Klafter J, Blumen A (1985) Chem Phys Lett 119: 377
72. Noolandi J (1977) Phys Rev B 16: 4466
73. Tsutsui T, Hatakeyama A, Saito S (1986) Chem Phys Lett 132: 563
74. Richert R, Bassler H (1985) Chem Phys Lett 116: 302
75. Richert R (1985) Chem Phys Lett 118: 534
76. Guillet J (1985) Polymer Photophysics and Photochemistry, Cambridge Univ. Press, Cambridge
77. Birks JB (1970) Photophysics of Aromatic Molecules, Wiley-Interscience, New York
78. Turro NJ (1978) Modern Molecular Photochemistry, Benjamin, Menlo Park
79. Terenin AN, Ermolaev VL (1952) Dokl Akad Nauk SSSR 85: 547
80. Terenin AN, Ermolaev VL (1956) Trans Faraday Soc 52: 1042
81. Dexter DL (1953) J Chem Phys 21: 836
82. Stern O, Volmer M (1919) Physik Z 20: 83
83. Perrin F (1924) Compt Rend 178: 1978
84. Inokuti M, Hirayama F (1965) J Chem Phys 43: 1978
85. Kobashi H, Morita T, Mataga N (1973) Chem Phys Lett 20: 376
86. Oster G, Geacintov N, Khan AU (1962) Nature 196: 1089
87. El-Sayed FE, MacCallum JR, Pomery PJ, Shepherd TM (1979) J Chem Soc Faraday Trans II 75: 79
88. Melhuish WH, Hardwick R (1962) Trans Faraday Soc 58: 1908
89. Jones PF, Siegel S (1969) J Chem Phys 50: 1134
90. Graves WE, Hofeldt RH, McGlynn SP (1972) J Chem Phys 56: 1309
91. Jassim AN, MacCallum JR, Moran KT (1983) Eur Polym J 19: 909
92. Horie K, Mita I (1982) Chem Phys Lett 93: 61
93. Horie K, Mita I (1984) Eur Polym J 20: 1037
94. Horie K, Tsukamoto M, Mita I (1984) Prepr. 1st SPSJ IPC: 96

95. Murov SL (1973) Handbook of Photochemistry, Marcel Dekker, New York, p 27
96. Horie K, Morishita K, Mita I (1983) Kobunshi Ronbunshu, 40: 217
97. Mita I, Takagi T, Horie K, Shindo Y (1984) Macromolecules 17: 2256
98. Noyes RM (1961) Prog React Kinet 1: 129
99. Vrentas JS, Duda JL (1977) J Polym Sci Polym Phys Ed 15: 417
100. Salmassi A, Schnabel W (1984) Polym Photochem 5: 215
101. Fraser IM, MacCallum JR, Moran KT (1984) Eur Polym J 20: 425
102. Gardlund ZG, Laverty ZG (1969) J Polym Sci, Polym Lett Ed 7: 719
103. Bertelson RC (1971) Photochromism. Brown GH (ed), Wiley-Interscience, New York, Chap 3
104. Richert R, Macromolecules, to be published
105. Hayashida S, Sato H, Sugawara S (1986) Polym J 18: 227
106. Ando E, Miyazaki J, Morimoto K, Nakahara H, Fukuda K (1985) Thin Solid Films 133: 21
107. Goldburt E, Shvartsman F, Fishman S, Krongauz V (1984) Macromolecules 17: 1225
108. Goldburt E, Shvartsman F, Krongauz V (1984) Macromolecules 17: 1876
109. Kalisky Y, Williams DJ (1984) Macromolecules 17: 292
110. Vilanove R, Hervet H, Gruler H, Rondelez F (1983) Macromolecules 16: 825
111. Suzuki Y, Ozawa K, Hosoki A, Ichimura K (1987) Polym Bull 17: 285
112. Munakata Y, Hatakeyama A, Tsutsui T, Saito S (1987) Rep Progr Polym Phys Jpn 30: (in press)
113. Heller HG, Szewczyk M: J Chem Soc Perkin I 1974: 1487
114. Heller HG, Langan JR: J Chem Soc Perkin I 1981: 341
115. Heller HG (1983) IEE Proc 130 Pt 1: 209
116. Murakami S, Tsutsui T, Tanaka R, Saito S: Nippon Kagaku Kaishi 1985: 1598
117. Murakami S, Tsusui T, Saito S, Yamato T, Tashiro M (1986) Rep Progr Polym Phys Jpn 29: 541
118. Priest W, Sifain MM (1971) J Polym Sci A-1, 9: 3161
119. Yu WC, Sung CSP, Robertson PE (1988) Macromolecules 21: 355
120. Yu WC, Sung CSP (1988) Macromolecules 22: 365
121. Horie K, Hirao K, Mita I (1986) Polym Prepr Jpn 35: 3234
122. Naito T, Horie K, Mita I (1987) Polym Prepr Jpn 36: 3575
123. Wendorff JH, Fischer EW (1973) Kolloid-Z, Z Polymere 251: 876
124. Roe RJ, Curro JJ (1983) Macromolecules 16: 428
125. Malhortra BD, Pethrick RA (1983) Eur Polym J 19: 457
126. Friedrich J, Haarer D (1984) Angew. Chem Int Ed Engl 23: 113
127. Moerner WE (1985) J Mol Electr 1: 55
128. Horie K, Hirao K, Kuroki K, Naito T, Mita I (1987) J Fac Eng, Univ Tokyo 39: 51
129. Thijssen HPH, Van der Berg RE, Volker S (1983) Chem Phys Lett 103: 23
130. Kador L, Schulte G, Haarer D (1986) J Phys Chem 90: 1264
131. Gutierrez AR, Friedrich J, Haarer D, Wolfrum H (1982) IBM J Res Develop 26: 198
132. Tani T, Namikawa H, Arai K, Makishima A (1985) J Appl Phys 58: 3559
133. Horie K, Kuroki K, Naito T, Mita I, Hirao K (1987) Polym Prepr Jpn 36: 3510
134. Winnacker A, Shelby RM, Macfarlane RM (1985) Opt Lett 10: 350
135. Lee HWH, Gehrtz M, Marinero EE, Moerner WE (1985) Chem Phys Lett 118: 611
136. Iannone M, Scott GW, Brinza D, Coulter DR (1986) J Chem Phys 85: 4863
137. Moerner WE, Carter TP, Brauchle C (1987) Appl Phys Lett 50: 430
138. Moerner WE, Levenson MD (1985) J Opt Soc Am B 2: 915
139. Lenth W, Moerner WE (1986) Optics Commun 58: 249
140. Cohen MD, Schmidt GMJ: J Chem Soc 1964: 1996, 2000
141. Dilling WL (1983) Chem Rev 83: 1
142. Bassler H (1984) Adv Polym Sci 63: 1
143. Sixl H (1984) Adv Polym Sci 63: 49
144. Enkelmann V (1984) Adv Polym Sci 63: 91
145. Minsk LM, Smith JG, Van Deusen WP, Wright JF (1959) J Appl Polym Sci 2: 302
146. Reiser A, Egerton PL (1979) Photogr Sci Eng 23: 144
147. Nakamura K, Kikuchi S (1967) Bull Chem Soc Jpn 40: 2684
148. Sitek F, Guillet JE, Heskins M (1976) J Polym Sci, Symp, 57: 343
149. Brandrup J, Immergut EH (ed) (1975) Polymer Handbook, 2nd edn, John Wiley, New York p V-13

150. Reich L, Stivala SS (1971) Elements of Polymer Degradation, McGraw-Hill, New York
151. Ranby B, Rabek JF (1975) Photodegradation, Photo-oxidation, and Photostabilization of Polymers, Wiley-Interscience, London
152. Jellinek HHG (ed) (1978) Aspects of Degradation and Stabilization of Polymers, Elsevier, Amsterdam
153. Schnabel W (1981) Polymer Degradation, Hanser, Munchen
154. Jellinek HHG (ed) (1983) Degradation and Stabilization of Polymers, Vol 1, Elsevier, Amsterdam
155. Mita I in Ref 152, Chapter 6
156. Charlesby A, Pinner SH (1959) Proc Roy Soc A, 249: 367
157. David C, Baeyens-Volant D (1978) Eur Polym J 14: 29
158. Geuskens G, Borsu M, David C (1972) Eur Polym J 8: 1347
159. Bowmer TN, O'Donnell JH, Winzor DJ (1981) J Polym Sci Polym Chem Ed 19: 1167
160. Wundrich K (1974) Eur Polym J 10: 341
161. Grassie N, Scotney A, Davis TI (1975) Makromol Chem 176: 963
162. Gesner BD, Kelleher PG (1969) J Appl Polym Sci 13: 2183
163. Carlsson DJ, Gan LH, Wiles DM (1978) J Polym Sci Polym Chem Ed 16: 2353
164. Obata K, Horie K, Mita I (1986) Disc Meeting Polym Degr Stabil, Prepr, Tokyo p 15
165. Decker C, Mayo FR, Richardson H (1973) J Polym Sci Polym Chem Ed 11: 2879
166. Decker C (1977) J Polym Sci Polym Chem Ed 15: 781
167. Decker C (1977) Makromol Chem 178: 2969
168. Jellinek HHG, Martuscelli E (1976) J Polym Sci Polym Phys Ed 14: 1249
169. Marchetti A, Martuscalli E (1976) J Polym Sci Polym Phys Ed 14: 151
170. Patel GN, Keller HH (1975) J Polym Sci Polym Phys Ed 13: 303, 322, 333
171. Patel GN (1975) J Polym Sci Polym Phys Ed 13: 339
172. Patel GN (1975) J Polym Sci Polym Phys Ed 13: 351
173. Sliinskas JA, Guillet JE (1973) J Polym Sci Polym Chem Ed 11: 3043
174. Shimada S, Hori Y, Kashiwabara H (1977) Polymer 18: 25

Editor: H. Fujita
Received February 19, 1988

Thermotropic Mesophases in Element-Organic Polymers

Yu. K. Godovsky, V. S. Papkov
Karpov Institute of Physical Chemistry, Institute of Synthetic Polymer Materials, USSR Academy of Sciences, Moscow

This review is primarily concerned with mesomorphic behaviour of element-organic polymers, such as linear and cyclolinear polyorganosiloxanes and linear polyphosphazenes. These polymers can form thermotropic mesophases without any mesogens. Recent development in thermodynamics, kinetics, structure and morphology, elasticity and thermoelasticity and some other properties are considered. The influence of the molecular structure and molecular weight of linear macromolecules on the ability to form the mesophases and their temperature stability is discussed. The phase behaviour of cyclolinear polyorganosiloxanes is described with special emphases on the temperature stability of the mesophases as a function of molecular structure, tacticity, substituents and flexible spacers. The state of order of polyorganosiloxanes is analysed in comparison to mesophase states of linear and cyclic low molecular siloxanes. The lyotropic mesophase behaviour of the ladder polyorganosiloxanes is also examined briefly.

List of Symbols and Abbreviations 131

1 Introduction. Classification of the Mesomorphic States 131

2 Mesophases in Polyorganosiloxanes 134
   2.1 Mesophases in Linear Polyorganosiloxanes 134
      2.1.1 General Consideration 134
      2.1.2 Phase Transitions and the Structure of the Crystalline Phases . . 135
      2.1.3 Structure and Morphology 139
      2.1.4 Kinetics of Formation 143
      2.1.5 Effect of the Molecular Mass on Thermodynamic Characteristics . . 146
      2.1.6 Effect of the Mesophase on the Formation of the Crystalline State 148
      2.1.7 Elasticity and Thermoelasticity 150
      2.1.8 Phase Behaviour of the Mixtures of Polydiethylsiloxane with
           Oligomers of Dimethylsiloxanes 153
   2.2 Cyclolinear Polyorganosiloxanes. 154
      2.2.1 General Description 154
      2.2.2 Phases and Phase Transitions 155
      2.2.3 Effect of Substituents 157
      2.2.4 Effect of the Chain Tacticity and Flexible Spacers 158
   2.3 Ladder Polyorganosiloxanes 160
   2.4 Low Mulecular Weight Siloxanes 164

3 Mesophases in Polyphosphazenes 167
   3.1 General Description 167
   3.2 Transitions 167
   3.3 Structure of Crystalline Phases and Mesophases 170
   3.4 Morphology of Crystalline and Mesomorphic State 174

4 Concluding Remarks 176

5 References 177

List of Symbols and Abbreviations

b_0	thickness of nucleus
C_1, C_2	constants in the Mooney-Rivlin equation
E	activation energy
f	retractive force
f_u	energy component of retractive force
G	free energy
H_i	enthalpy of isotropization
H_m	enthalpy of fusion
K	overall rate of a transition
M	molecular weight (mass)
n	Avrami parameter
Q	heat
$\langle r_0^2 \rangle$	mean square end-to-end distance of unperturbed chains
T	absolute temperature
T_g	glass transition temperature
T_i	isotropization temperature
T_i^0	isotropization temperature of lamellae of infinite thickness
T_m	melting point
U	internal energy
V	linear growth rate; volume
W	work
λ	elongation ratio
σ	interfacial free energy parallel to the molecular chain direction
σ_e	interfacial free energy perpendicular to the molecular chain direction
PDPhS	polydiphenylsiloxane
CLPOS	cyclolinear polyorganosiloxanes
POSSO	polyorganosilsesquioxane
OPCTS	octaphenylcyclotetrasiloxane
DOTES	1,3-dioxytetramethyldisiloxane
DOTPS	1,3-dioxytetra-n-propyldisiloxane
DECPS	decaethylcyclopentasiloxane
PBFP	poly bis(trifluoroethoxy)phosphazene
PDES	polydiethylsiloxane
PDPS	polydipropylsiloxane
PDMS	polydimethylsiloxane

1 Introduction. Classification of the Mesomorphic States

This review concerns the mesomorphic state of the element-organic polymers. In this connection, a question arises, what is the reason for separating the mesophase element-organic polymers into a special class. To answer this question, it is necessary to consider the contemporary classification of the mesomorphic states in general and mesomorphic states of polymers in particular.

According to the modern classification of the mesophase states, three various types of mesophases, viz., the thermodynamically stable phases intermediate between the crystalline and liquid phases, can exist: liquid crystals, plastic crystals and condis crystals [1,2]. The liquid crystal mesophase is the most thoroughly studied mesophase at present. Many classes of small organic molecules can form liquid crystalline phases and the main conditions for the occurrence of this state in low molecular substances are the asymmetrical shape of the molecules and their rigidity. Such molecules or fragments are called mesogens. The liquid crystalline state is characterized by the absence of the positional long range order in one dimension. The remaining positional and/or long range orientational order gives rise to the anisotropic physical properties of the liquid crystalline state while the absence of this positional long range order is the basis for the liquid properties. Therefore, the liquid crystals are considered as positionally disordered crystals or orientationally ordered liquids [2].

Recent studies have demonstrated that in the liquid crystalline state not only small molecules but also polymers can exist. For the first time the liquid crystalline state in polymers has been discovered in solutions of rigid rod chain polymers [3,4]. These rigid rod macromolecules cannot be transformed from the solid into the liquid state without degradation and, therefore, they require the presence of a solvent in order to produce a mesophase (lyotropic mesophase). To produce the polymers which exhibit a liquid crystalline state at temperatures above the crystalline solid state and below the transformation into an isotropic melt (thermotropic polymers) the obvious idea to tie up the mesogenic molecules to macromolecules was successfully used [5-7]. Such thermotropic mesophase-forming polymers contain either the mesogenic groups directly in the main chain (liquid crystal main chain polymers), or the mesogens may be linked as side chains to the macromolecules (side chain liquid crystalline polymers). While for the liquid crystal main chain polymers, more or less the macromolecule as a whole takes part in the formation of the liquid crystalline structure; for the liquid crystalline side chain polymers the anisotropic arrangement is built up only by mesogenic side chains without anisotropic packing of the backbone. Although the linkage to the main chain restricts rotational and translational motions of the mesogenic side chains, the liquid crystal structure for these polymers is very similar to that of small molecules.

It is very important to emphasize that, without any mesogens incorporated in macromolecules, the formation of a thermotropic liquid crystal structure in polymers is unlikely [2]. Nevertheless, there is a point of view in polymer literature that in the melt of flexible polymers a liquid crystal type structure may exist [8-10]. This conclusion is based on the fact that asymmetric shape is the immanent characteristic of any macromolecule. This approach totally ignores the role of the rigidity which is also a very important characteristic of macromolecules. For rigid, rodlike, liquid-crystal polymers, the Kuhn segment reaches hundreds and even thousands of angstroms [11] and for such polymers "asymmetry of the chemical structure corresponds to the geometrical anisotropy of the macromolecules" [3]. For flexible macromolecules the lengths of the Kuhn segments are 1 to 30 Å. According to Flory, [4] such values of Kuhn segments cannot stimulate the occurrence of the liquid crystalline state. Therefore, above the melting point in the melt of flexible macromolecules, only the short range order similar to that of small molecules can exist. The macroconformation of flexible polymers in the melt is similar to that in θ-solvents. Recent studies of the order in the melt of flexible macromolecules, in particular, with the aid of SANS demonstrated that the orientational correlation along the macromolecules exists only at the distances of 1 to 30 Å, i.e. it is determined by the Kuhn segment in full accord with Flory's prediction. Hence, the short range order in the melt of flexible polymers such as polyethylene and polypropylene cannot be treated as a special liquid crystalline state.

Some small molecules possessing a spherical (or close to spherical) structure such as N_2, CCl_4, CH_4, SiH_4, and some others can exist in the mesomorphic state known as the plastic crystal [1, 2]. This mesophase state is characterized by the positional order of the centres of gravity of molecules with an absence of orientational order. Although there are attempts in the literature to consider some mesophases in polymers as the state of plastic crystal Wunderlich and Grebowicz [2] concluded recently that this mesophase state really exists only in the case of small molecules.

In the last decade, it has been established that some flexible macromolecules without any mesogens in their chemical structure are able to form thermodynamically stable thermotropic ordered phases possessing the structure and properties intermediate between the crystalline and amorphous phases. It is important that such mesophases exist without the action of any external orientational forces. For the first time, such mesophases were discovered in polymers with inorganic backbone polyphosphazenes and polysiloxanes. Desper and Schneider [12] have found that upon heating some crystalline polyphosphazenes are transformed into a mesomorphic phase. Later, a similar mesophase was found in linear polydiethylsiloxanes by Beatty and co-workers [13]. For this mesophase they used the term "viscocrystalline". Having analyzed the structure of these mesomorphic phases, Schneider et al. [14] have suggested that the mesophases can be considered as disordered crystals and, therefore, one can attempt to find analogous phases among the disordered organic crystals. Extending this idea and the classification of the mesophase states given by Smith [1], Wunderlich and Grebowicz [2] proposed to consider the additional type of the mesomorphic state — the conformationally disordered crystals (condis crystals). This term was introduced to designate this new type of the mesophase — the most important mesophase for flexible linear macromolecules as has been emphasized by the authors. "The condis crystals consist of flexible molecules which can undergo relatively easily hindered rotation to change conformation without losing positional or orientational order" [2].

Although the crystalline form of polytetrafluoroethylene existing between 303 K and the melting point (600 K), the high temperature hexagonal crystalline modification in 1,4-trans-polybutadiene, the high pressure phase of polyethylene and some other crystalline forms of various polymers have also been considered lately as condis crystals [2]. In a most striking form, the tendency to build an ordered state in polymers without mesogens, is displayed by the already mentioned element-organic macromolecules. This was the first reason for considering the mesomorphic state of these polymers separately.

In the 1980's, a novel class of mesomorphic polysiloxanes has been added to the family of element-organic mesomorphic macromolecules viz. the cyclolinear polyorganosiloxanes. Depending on the nature of substituents and the type of linking of cycles in macromolecules, the flexibility of such cyclolinear macromolecules can change considerably from being very flexible, similar to typical linear polyorganosiloxanes, to relatively rigid. This gives the opportunity of elucidating the role of chain flexibility (rigidity) in the formation of the thermotropic mesophase state. It was the second reason for separate consideration of the mesomorphic state of element-organic polymers.

Thus, the subject of the review are linear and cyclolinear element-organic polymers forming a specific type of the mesophase state. The structure of the mesophases and the formation of liquid crystalline macromolecules, in which classical mesogens in the main or side chains are linked by flexible polysiloxane [6,7] or polyphosphazene spacers [15-17], are similar to the liquid-crystalline polymers with organic spacers. Structure and properties of these polymers have been reviewed recently [6,7] and, therefore, will not be considered in our article.

2 Mesophases in Polyorganosiloxanes

2.1 Mesophases in Linear Polyorganosiloxanes

2.1.1 General Consideration

Linear polysiloxanes have the general chemical structure

$$\left[\begin{array}{c} R \\ -Si-O- \\ R \end{array}\right]_n$$

where R are various alkyl and aryl substituents. Polysiloxane macromolecules are very flexible and the rotational energy is of the order of kT [18]. The rigidity of siloxane macromolecules is determined by steric interaction of their side groups. The most flexible macromolecules belong to PDMS with $(\langle r^2 \rangle / \langle r_0^2 \rangle)^{1/2} = 1.6$ [19]. With increasing length of the alkyl substituents the rigidity of the chain increases slightly: PDES $(\langle r^2 \rangle / \langle r_0^2 \rangle)^{1/2} = 1.75$ [20], PDPS $(\langle r^2 \rangle / \langle r_0^2 \rangle)^{1/2} = 1.9$ to 2.0 [19]. In spite of the slight increase of the chain rigidity with length of the side group, all polyalkylsiloxane macromolecules are flexible. At present there are no characteristics of the flexibility of polyarylsiloxanes, in particular, polydiphenylsiloxane.

Although linear polyorganosiloxanes have been known for a rather long time, their structure and properties were extensively studied only for PDMS because of its industrial importance. Uncrosslinked linear PDMS is a relatively fast-crystallizing polymer with $T_g = 150$ K and $T_m = 235$ K, i.e. far below room temperature. Above T_m, the PDMS is in an amorphous liquid-like state and there is no information in the literature concerning the possibility of the existence of any mesomorphic state. The first information about the phase states and transitions in PDES and PDPS was published by Lee et al. [21]. They discovered low temperature solid state transitions in these polysiloxanes but no mesomorphic states were found. In a more thorough study of PDES [13], a mesomorphic state was found after melting. The mesophase existed in the temperature interval 270 to 300 K, but above 300 K it was transformed into the fully amorphous melt. The mesophase state was called "viscocrystalline" and the authors made an attempt to characterize it with scanning calorimetry, NMR, dielectric and optical methods. But neither the structure, nor the physical properties of the mesomorphic state were studied completely enough. A mesomorphic state in PDPS was first discovered and studied in our investigations [22-25]. Any information about the mesomorphic state of other representatives of linear polyalkylsiloxanes is absent today. Recently, we found that linear PDPhS was also transformed into the mesomorphic state after melting [26]. The transformation into the amorphous molten state was not reached because of its chemical degradation upon heating to high temperatures. Hence, at least three linear polysiloxanes can exist in the mesomorphic state.

2.1.2 Phase Transitions and the Structure of Crystalline Phases

A characteristic feature of linear PDES and PDPS, besides their ability to exist in the mesomorphic state, are the temperature transitions in the crystalline state at low temperatures [22,23,27-29] (Fig. 1). Four temperature transitions occur on heating the samples from low temperatures: glass transition (for PDES of high crystallinity this transition at T_g is practically unrecognizable), a single or double endothermal peak in the vicinity of 190 to 220 K, resulting from the crystal-crystal transition, a single or double peak resulting from the melting of the crystalline structure and, finally, a small endothermal peak accompanying the transition from the mesomorphic to the isotropic liquid state. Thermodynamic characteristics of all these transitions are listed in Table 1. Data for PDMS and PDPhS are also included in the Table. The low temperature transitions are absent in these polymers. The total heat of all the phase transitions involved in transforming PDES and PDPS from the crystal-

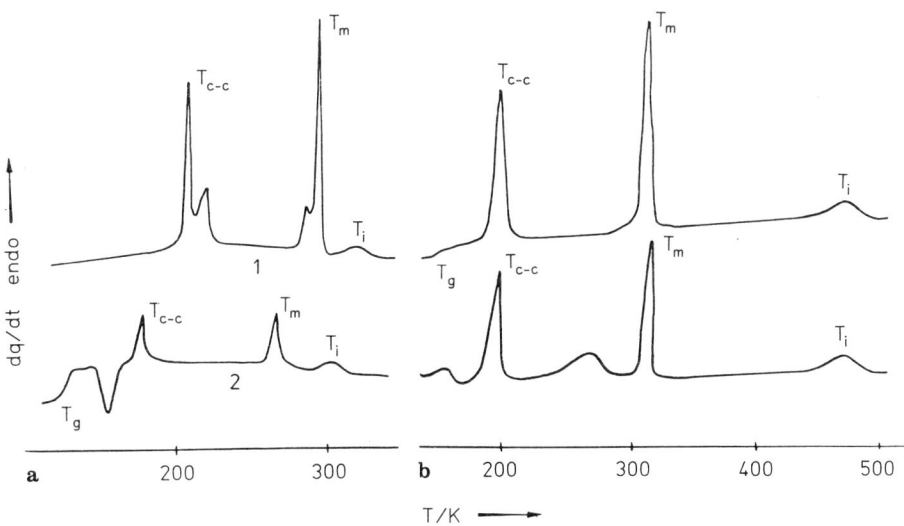

Fig. 1a and b. DSC records of PDES and PDPS samples with various thermal histories [22]. **a** PDES, **b** PDPS. 1 — slow cooling from room temperature to 150 K; 2 — quenching from above T_i to the liquid nitrogen temperature. Heating rate — 20 K/min. Values of T_g, T_{c-c}, T_m and T_i are listed in Table 1

Table 1. Thermal transition data for linear polyorganosiloxanes [25, 27]

Polymer	Thermal history	T_g K	Crystal-Crystal		Melting		Isotropization		Crystallinity, %
			T_{c-c}, K	ΔH_{c-c}, J/g	T_m, K	ΔH_m, J/g	T_i, K	ΔH_i, J/g	
PDMS	Slow cooling	150	—	—	238	35	—	—	55
PDES	Slow cooling	—	214	28.0	280	17.0	319	3.0	90
α-form	Quenching	134	180	4.2	263	4.2	307	3.0	20
β-form	Slow cooling	134	206	26.0	290	21.0	319	3.0	90
PDPS	Slow cooling	164	218	15.2	333	22.0	479	3.2	70
	Quenching	164	216	10.7	333	12.8	479	3.0	43
PDPhS	Slow cooling	—	—	—	545	35.5	—	—	—

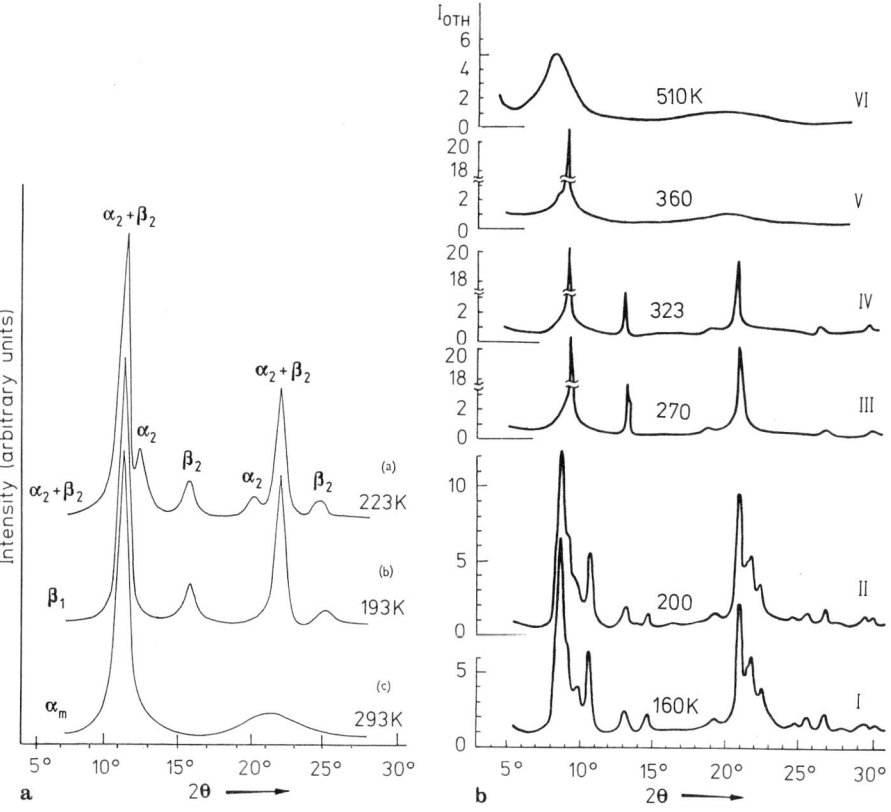

Fig. 2a and b. Diffractograms of isotropic PDES and PDPS at different temperatures [23, 30]. **a** PDES, **b** PDPS, I — 160 K, II — 200 K, III — 270 K, IV — 323 K, V — 360 K, VI — 510 K

line into the amorphous state is approx. 5 kJ/mole per repeating unit and it is close to the heat of fusion of the completely crystalline PDMS (4.7 kJ/mole), which is not able to form a mesophase state. This seems to mean that the contribution of the conformations to the crystal lattice energy for linear polyalkylsiloxanes is also close. The heat of the transition to the isotropic state is about 15% of the heat of fusion of the crystalline phase which considerably exceeds the amount characteristic of the mesogenic polymers. However, this relation for polysiloxanes decreases some percent if the heat of low temperature transitions is also taken into account. The T_g/T_m for PDES and PDPS is close to 0.5 but differs from PDMS with T_g/T_m equal to 0.64 and PDPhS with T_g/T_m equal to 0.67.

X-ray studies prove the calorimetric identification of the phase transitions [30]. Typical changes of X-ray diffraction spectra with temperature are shown in Fig. 2. These data confirm the existence of the low temperature crystal-crystal transition

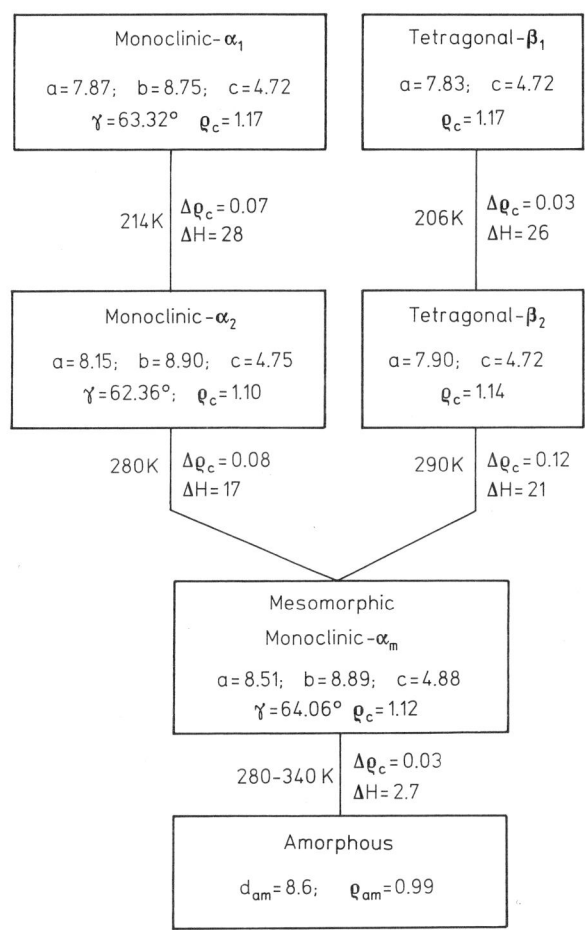

Fig. 3. General diagram of the structural changes accompanying the phase transitions in PDES [30]. The cell parameters are in Å, ΔH in J/g and ϱ_g and $\Delta\varrho_c$ in g/cm^3

as well as the transition of the high temperature crystal polymorphs into the mesomorphic state. The latter transition is accompanied by the disappearance of the crystalline reflections apart from the only reflection in the range of $2\theta = 10°$ corresponding to the distance between the macromolecular backbones. Finally, the isotropization which occurs on further heating the mesophase is accompanied with the disappearance of the single sharp reflection and with the formation of a diffraction profile spectrum typical of amorphous polysiloxanes.

The general diagram of the structural changes accompanying the phase transitions in PDES is shown in Fig. 3 [27,30]. The low-temperature transitions are featured by the fact that the type of the crystal lattice remains the same under jump-like changes of its parameters and crystalline density. The density of the amorphous and mesomorphic states are almost the same and, therefore, there are neither long periods nor small-angle scattering in the mesomorphic state, although in the crystalline state, both for the oriented and isotropic samples, the long period 490 Å practically independent of temperature exists. The structure of the mesophase will be discussed later.

It is necessary to note that $\alpha_1 \rightarrow \alpha_2$ and $\alpha_2-\alpha_{mes}$ transitions may be considered as transitions of the same type since the crystalline symmetry remains the same and only the parameters of crystalline lattices become larger. This fact distinguishes the PDES mesophase from those of polyphosphazenes for which pseudohexagonal packing of chains is normally observed. However, the difference between the monoclinic stucture of the mesophase of PDES and the hexagonal packing is very small. The identity period in the crystalline state is 4.72 to 4.75 Å which corresponds to two $(C_2H_5)_2SiO$ units per single period. The conformation of the PDES macromolecule in the crystalline state is not yet clear, but it must differ slightly from the plane-extended cis-trans conformation with the identity period of approx. 5 Å.

Table 2. Pulse NMR data on molecular dynamics in various phases of PDES [31]

Phase	Temperature range K	Motion modes and frequencies		
		C_3-rotation of methyl groups	Rotation of ethyl groups around Si—C bond	Chain dynamics
β_1	Below 200	$v_c \leq 10^9$	$v_c \; 10^4$	—
α_1	Below 210	$v_c \cong 10^8$	—	—
β_2	200–283	$10^9 \leq v_c \leq 6 \cdot 10^9$	$10^9 \leq v_c \leq 10^{10}$	$> 10^8$ $\sim 10^8$ at 283 K $\}$ swinging
α_2	210–273	$10^9 \leq v_c \leq 6 \cdot 10^9$	$10^9 \leq v_c \leq 10^{10}$	$> 10^8$ $\sim 4 \cdot 10^3$ at 248 K $\}$ swinging $\sim 10^8$ at 260 K
α_m	273–305	$v_c \cong 8 \cdot 10^9$	$v_c \cong 10^{10}$	$\sim 10^3$ at 278 K $\}$ chain $\sim 10^8$ at 293 K $\}$ bending $\sim 10^9$–10^{10} chain units rotation (conformational motion)

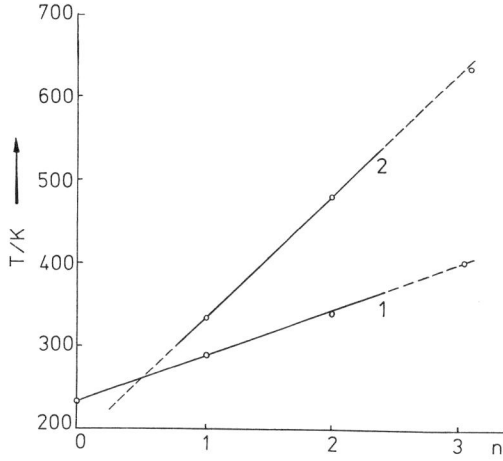

Fig. 4. Dependence of melting point and isotropization temperature on number of CH_2-groups in the repeating unit of polyorganosiloxane $\{Si[(CH_2)_nCH_3]_2\}_p$ [23]. 1 — melting point, 2 — isotropization temperature. n = 0 — PDMS, n = 1 — PDES, n = 2 — PDPS, n = 3 — predicted values for poly(dibutylsiloxane)

The phase transitions in PDES are accompanied with jump-like changes of the molecular motion, as seen from Table 2, where the data or pulse NMR concering the type and frequencies of the molecular motion are listed [31]. The data show indirectly that a considerable conformational disordering takes place not only at the crystal-mesophase transition but also at the low-temperature solid-solid transitions.

For PDPS, such complete information concerning the structure of crystalline phases and molecular motion is not available at present.

To conclude the general description of the temperature transitions in linear polysiloxanes, it is appropriate to consider the dependence of the temperature transitions on the chemical structure of the siloxane chain (Fig. 4). It is seen that the melting point depends linearly on the number of CH_2-groups of side chains. poly(dibutylsiloxane) leads to T_m equal to 395 K. If the dependence of T_i on n is also linear, its slope should be considerably larger. Extrapolation of T to PDMS and poly(dibutilsiloxane) — leads to 180 K and 600 K, respectively. The predicted value for poly(dibutylsiloxane) cannot be checked at present for lack of this polymer. The predicted value of T_i for PDMS should be considerably below T_m. This fact seems to be a formal reason of the absence of the mesomorphic state in PDMS.

2.1.3 Structure and Morphology

A characteristic feature of the diffractograms of the mesophases in PDES and PDPS consists in the existence of two maxima (Fig. 2). One of them (rather narrow and very intensive) is near $2\theta = 8$ to $12°$, and the second one with a wide maximum is near $2\theta = 16$ to $25°$. The fact that there is one sharp X-ray maximum supported by optical and NMR data, has led Beatty et al. [13] to the conclusion that PDES can exist in the mesomorphic state. The structure of the mesophase was identified by Tsvankin et al. [30] who obtained a well-oriented mesomorphic structure by stretching a slightly crosslinked PDES film five to six times its original length at room temperature (Fig. 5). The X-ray pattern is typical of mesomorphic structures which are intermediate between crystalline and amorphous structures. On the equator (see also Fig. 2), the main reflection is at $2\theta = 11.03°$, $d = 8.01$ Å and

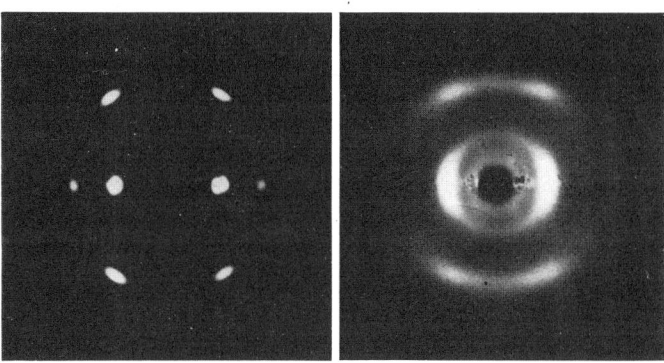

Fig. 5. X-ray fiber photographs of the oriented PDES film [30]. 1 — 193 K (crystalline state), 2 — 293 K (mesomorphic state)

two weak reflections with d_2 = 4.61 to 4.62 Å and d_3 = 3.04 to 3.05 Å. The first layer line retains a strong but also broad reflection with d = 4.17 Å. The identity period determined from this reflection is 4.90 Å. What distinguishes the X-ray photograph of a mesomorphic structure is a rapid decrease in intensity with increasing 2θ, with the result that there are no further reflections on the zeroth and first layer lines. The sharp decrease of intensity is a consequence of the disordering of the chains, both rotational and translational, along the chain axes. The considerable broadening of the reflection on the first layer line indicates that in the mesomorphic structure there probably exists a variety of conformations with the identity periods of 4.85 to 4.95 Å. Although the equatorial reflections of the mesomorphic structure are in a fairly good agreement with those of a hexagonal structure, the theoretical density of the structure (0.936 g/cm^3) is considerably less than the experimentally determined density of non-crosslinked PDES in the mesomorphic state at room temperature (1.015 g/cm^3). The hexagonal unit cell hypothesis was, therefore, discarded. A thorough analysis has shown that the mesomorphic structure corresponds most probably to the monoclinic crystalline lattice with a = 8.51, b = 8.89 and c = 4.88 Å and γ = 64.06°. The density of the lattice is equal to 1.02 g/cm^3 which is close to the experimental value. Thus, the mesomorphic structure consists of the monoclinic unit cells which have parameters close to those of the monoclinic unit cells α_1 and α_2 (see Fig. 2). The higher values of the parameters of the mesomorphic monoclinic unit cell and its lower density compared to α_1 and α_2 are sufficient to cause rotational and translational disordering of the chains. Such disordering manifests itself as a sharp decrease of intensity in the X-ray patterns of the mesomorphic structure. However, the actual structure distortion is apparently insufficient for a hexagonal packing. A hexagonal unit cell might be formed in case of a cylindrical symmetry of the macromolecules and free rotation of the chains about their axes. In the case of PDES such free rotation would require too much free space, because the polymer chains contain longer ethyl side groups. Apparently, this is energetically unfavourable and the structure that is formed is not a hexagonal mesomorphic structure with complete rotational disorder but a monoclinic mesomorphic structure with a partial rotational disorder and higher density. This is a very good example of a condis crystal.

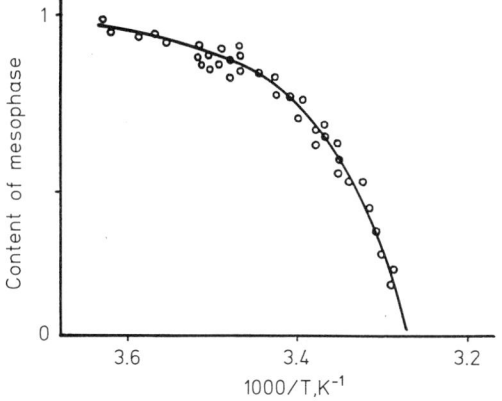

Fig. 6. Content of the mesophase in PDES sample as a function of reciprocal temperature [33]

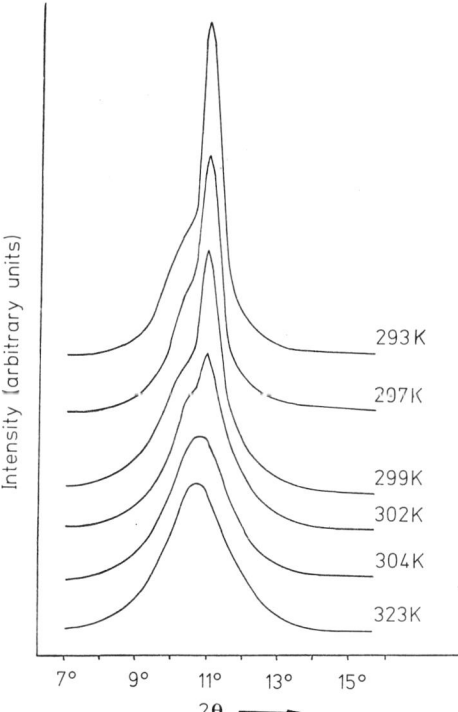

Fig. 7. Wide-angle difractograms of isotropic crosslinked PDES film at different temperatures illustrating the gradual transition from the mesomorphic to amorphous state [30]

Although X-ray fiber patterns of PDPS are not available yet and the mesophase structure has not been identified, one can suggest that the above concept is also true for PDPS.

An interesting feature of the mesomorphic state in PDES is the coexistence of the mesomorphic and amorphous phases. This conclusion follows both from X-ray [30] and NMR [32,33] studies. A typical dependence of the mesophase content on temperature obtained by transverse NMR-relaxation is shown in Fig. 6. It is important to emphasize that in the temperature region of the transition from the crystalline to the mesomorphic state the content of the mesophase is 0.95 which means that practically all the polymer is in the mesophase state. A decrease in the content of the mesophase with temperature can be easily determined by the X-ray and DSC analysis.

The mesomorphic state reflections are the strongest at temperatures at which the melting of the crystalline modifications is completed. On further increasing the temperature the reflection intensities decrease and in thermograms this decrease manifests itself as larger area of the endothermal peak of the transition from the mesomorphic to amorphous state. The temperature range, in which this transition occurs, depends on the sample type (molecular mass, degree of crosslinking, heat treatment history)[27,28]. Typical results are shown in Fig. 7. As the temperature increases the mesomorphic reflection ($2\theta = 11.03°$), the intensity decreases and the intensity of the amorphous halo ($2\theta = 10.65°$) increases. At 303 K, the sample becomes completely amorphous. The data obtained from analysis of the diffractograms and thermograms are in good agreement[30]. Rough estimates show that at room temperature the content of the mesomorphic phase in the investigated sample was about 30%. The amorphous halo is observed in all X-ray diffractograms at different temperatures. Its intensity depends on the temperature and sample type. In particular, the intensity of the amorphous halo of linear uncrosslinked PDES at low temperatures is small and, as we have already noted, its crystallinity must be very high (not less than 90%). Below 303 K, the position of the halo center corresponds to $d_{am} = 8.3$ Å. Above 303 K, when the structure becomes fully amorphous, the position of the halo center slightly changes: at 323 K, $d_{am} = 8.4$ Å. The broadening of the mesophase reflection and its transformation to the fully amorphous halo is accompanied with a complete optical isotropization. The density of the amorphous crosslinked film at 293 K is 0.99 g/cm^3, which is very close to the PDES density in the mesomorphic state. One may, therefore, assume that in the amorphous state the short range order remains similar to the ordering in the mesophase state. Finally, it is worth mentioning that the diffraction pattern of the isotropic amorphous PDES and PDPS differs from the corresponding pattern of the flexible carbon chain polymers: there are two amorphous halos, the first one being rather intensive. The intensity of the first reflection changes with temperature reversibly. Similar behaviour resulting from the thermal expansivity has been found recently for some other macromolecules with a inorganic backbone[26].

The morphology of mesomorphic PDES and PDPS differs from the morphology of liquid-crystalline polymers with traditional mesogenes which is, in general, similar to the morphology of low molecular weight liquid-crystalline substances. On the contrary, the morphology of PDES and PDPS resembles more the morphology of crystalline polymers[22,35]. The colorless transparent PDES melt becomes turbid during the mesophase appearence. Between the crossed polarizers, the formation of the mesophase can be observed as the appearance and development of light bands, the length and intensity of which increase in time. This is seen in Fig. 8 where some microphotos are presented, they were taken at consecutive stages of the mesophase formation. The intensity of these light bands is maximum if they incline at 45° to the vibration direction of the polarizer and they are extinguished when they are parallel or normal to the vibration direction of the polarizer. Comparison of the pictures taken by in reflected and transmitted light at the terminal stage of the

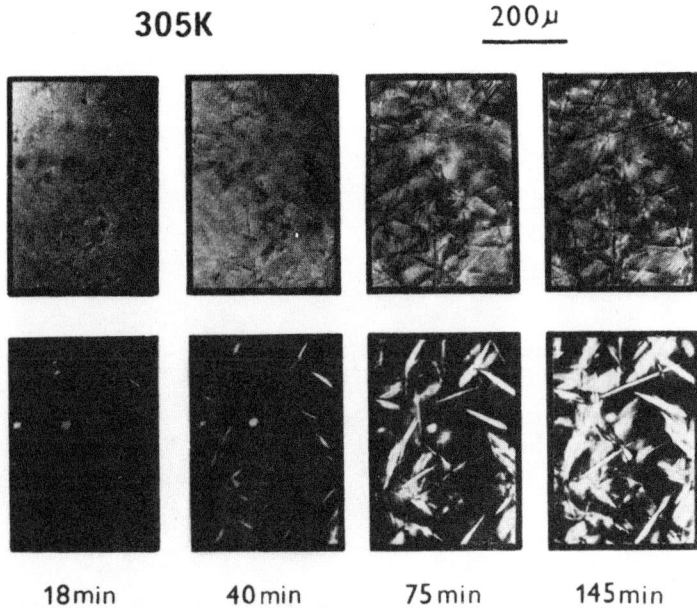

Fig. 8. Photomicrographs of a PDES sample in the transmitted light between crossed Nicol prisms (below) and the surface of the same sample in reflected light (above) at different stages of formation of the mesophase at 305 K [35]

mesophase formation leads to the conclusion that the light bands are apparently lateral faces of lamellar domains of the mesomorphic phase. Apart from the lamellae emerged on the surface of the sample, those embedded in the body are also seen. In reflected light, one can also see edges and faces of the lamellae inclined at small angles to the figure plane but between crossed Nicol prisms in transmitted light these lamellae are not distinguishable.

Further evidence of lamellar morphology was obtained from the measurement of the sign of birefringence of the light bands [35]. It was established that both single bands and the set of parallel bands show a neagtive birefringence. To correlate the sign of the birefringence with the direction of macromolecule axes in the bands, the birefringence of the crosslinked PDES film stretched five-fold) was determined. Such film exhibits well-defined fiber mesomorphic texture. Above 285 K, the refractive index is greater for light vibrating parallel to the extension direction, i.e. the film is positively birefringent. This means that the refractive index is greater for the direction parallel to macromolecules axes correspondingly macromolecules are arranged normally to the longitudinal axis of the light bands in the isotropic sample. Naturally, it is possible only in the case of lamellar morphology suggested above from a consideration of optical data.

2.1.4 Kinetics of formation

Information about the kinetics of mesophases formation in polymers is very scarce at present, although it is quite obvious that such a study may throw additional light

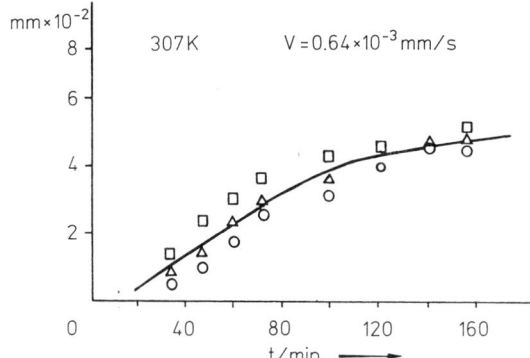

Fig. 9. The growth of the band length of PDES with time at 307 K [35]

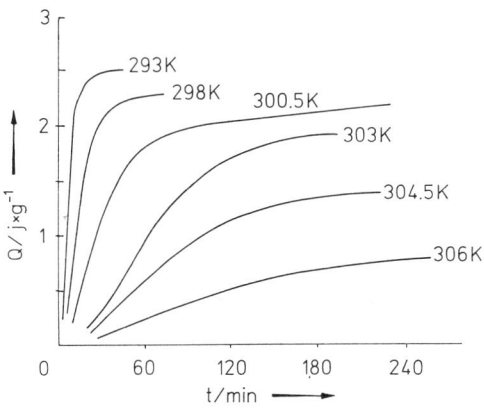

Fig. 10. Kinetics of the mesophase formation in PDES at different temperatures [35]

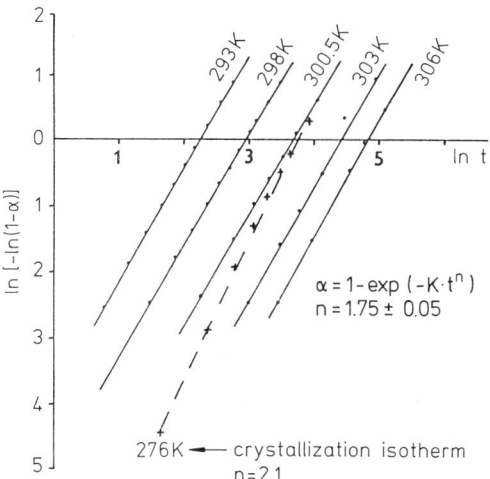

Fig. 11. Plot of $\ln[-\ln(1-\alpha)]$ vs $\ln t$ at different temperatures of the mesophase formation in PDES. The isotherm of crystallization at 276 K is also shown [22, 35]

on the nature of the mesomorphic states. A complex kinetic study of the formation of mesophase in PDES published recently [35] included the measurements of the linear growth rates of the lamella-type mesophase domains and the calorimetric overall kinetics of the mesophase formation.

It was established that the linear growth rate of the mesophase domains (the light bands) is constant during a rather long period of time and only at the terminal stages, a certain levelling-off occurs (typical results are shown in Fig. 9). The calorimetric isotherms of the heat evolution resulting from the mesophase formation (Fig. 10) have an S-shape as in the case of crystallization of polymers. However, the heat of transition decreases considerably when temperature increases.

The analysis of these calorimetric isotherms in terms of the Avrami equation (Fig. 11) shows that the kinetics of the mesophase formation is described by the Avrami parameter $n = 1.75 \pm 0.05$. Such a value of the morphological parameter was interpreted as two-dimensional growth of lamellar structures on heterogeneous nuclei.

Interesting conclusions were drawn from a comparison of the temperature dependence on the linear growth rate and overall rate of the mesophase formation. Assuming the growth of the mesophase domains is similar to that of polymer crystallites, their linear growth rate must be determined by the secondary nucleation and can be described by the following equations [36]

$$V = \text{const} \cdot \exp(-E/kT) \exp(-G/kT) \tag{1}$$

with

$$G = 4\sigma\sigma_e b_0 T_i^\circ/(\Delta H_i \Delta T) \tag{2}$$

where E is the activation energy for the transport process at the interface, G is the free energy of the critical size nucleus formation, b_0 is the thickness of the surface

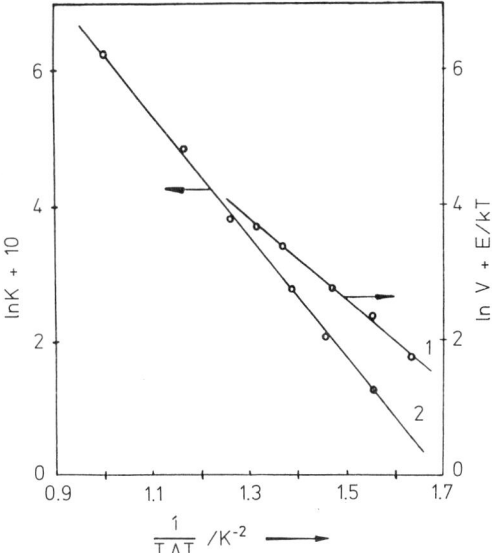

Fig. 12. Plots of $(\ln K + 10)$ and $(\ln V + E/kT)$ vs $1/(T \Delta T)$ for PDES [22, 35]

layer defined by the lattice parameter, σ and σ_e are the interfacial free energies per unit area parallel and perpendicular, respectively, to the molecular chain direction.

Figure 12 shows that the temperature dependence of the linear growth rate of the mesomorphic lamellae really follows Eqs. 1 and 2 with the slope $S_L = 5.8 \times 10^4 \, K^2$.

According to theory [36], the overall rate of the transition K may be expressed as follows

$$K = aBV^n \tag{3}$$

where a depends on the growth geometry, V is the linear growth rate, and B is the nucleation rate for the homogeneous nucleation or the number of heterogeneous nuclei in the corresponding space if the nucleation is a heterogeneous process. Figure 12 shows that the experimental data obtained follow Eq. (3) very well. The slope $S_A = 9.65 \times 10^4 \, K^2$ and the ratio $S_A/S_L = 1.67$ which is very close to the value of the morphological parameter $n = 1.75$. According to Eq. 3 such coincidence corresponds nominally to the formation of the mesophase on heterogenous nuclei, the number of which is constant in the whole temperature range studied. Taking into account the lamella-type morphology of the mesophase one can apparently speak about the two-dimensional growth of heterogeneous nuclei.

Thus, the morphology of the mesomorphic flexible PDES resembles in many respects the morphology of crystalline polymers. However, in contrast to the usual crystalline lamellae the mesophase lamellae have considerably larger dimensions, their thickness being 1.5 to 2 μ, they consist of some layers with extended-chain macromolecules.

2.1.5 Effect of the molecular mass on thermodynamic characteristics

Recently, it has been established that the mesophase state is much more sensitive to molecular mass changes than the crystalline state [5-7]. The behaviour of nonfrac-

Table 3. Influence of molecular weight on thermal characteristics of the mesophase of PDES [22, 25] and PDPS [25]

Polymer	$\bar{M}_n$ 10^{-3}	T_i K	ΔH_i kJ/mole	ΔS_i J/mole K
PDES	58	296	0.26	0.88
	100	307	0.27	0.88
	172	319	0.27	0.85
	425	325	0.28	0.86
	765	326	0.28	0.83
PDPS	8	—	—	—
	27	342	0.065	0.19
	36	395	0.25	0.38
	43	418	0.10	0.24
	51	445	0.15	0.30
	68	450	0.39	0.80
	82	479	0.39	0.80
	87	480	0.39	0.80

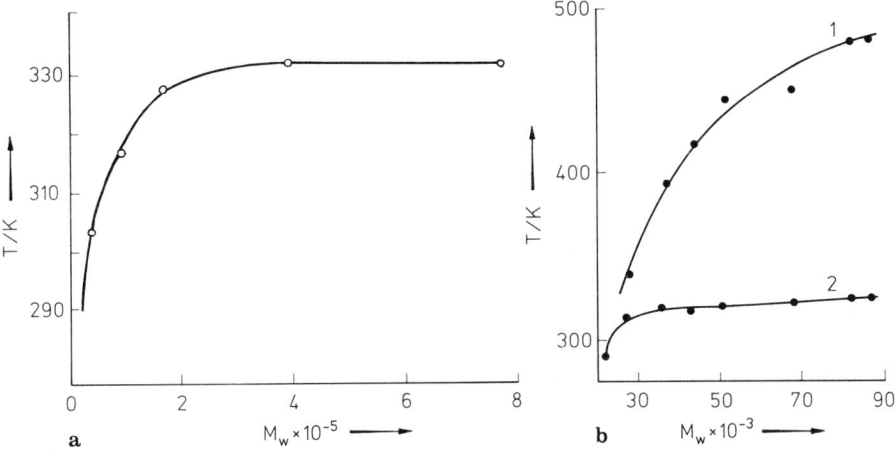

Fig. 13a and b. Dependence of the isotropization temperature and melting temperature on the molecular weight [22,23,27]. a PDES, b PDPS. 1 — T_m, 2 — T_i

tionated samples of PDES and fractionated samples of PDPS supports the strong influence of the molecular weight on the isotropization temperature (Table 3). A different dependence of T_m and T_i on the molecular weight leads to a considerable increase of the temperature interval of existence of the mesophase if the molecular weight increases [22,25,28,29,35] (Fig. 13). It is noteworthy that the lowest molecular weight PDPS fraction is not able to form the mesophase at all. Hence, the temperature interval of existence of the mesophase in PDPS is about 150 K and in PDES — about 50 K. Such a considerable increase in the temperature interval of the mesomorphic state on going from PDES to PDPS seems to be connected with an increased rigidity of the siloxane chain due to increased length of the side chain. Following the behaviour of crystalline polymers it was suggested that this dependence is either a result of the influence of end groups on T_i, or it reflects the influence of morphology on T_i. A check of the former suggestion for the fractions of PDPS showed that, although the dependence of $1/T_i$ on $1/M$ is linear, ΔH_i found from the slope of the line is about an order of magnitude less than the experimental value. Hence, the first suggestion can be rejected.

The second suggestion is based on morphological findings and kinetic studies of the growth of the mesophase lamellae and the overall kinetics of the transition isotropic liquid-mesophase. As has been mentioned above, the thickness of the mesophase lamellae is 1.5 to 2 μ. Taking into account a very high degree of crystallinity and the content of the mesophase close to 100%, one may suggest that the most reliable conformation of macromolecules is the extended chain conformation. The corresponding estimation for PDES shows that any lamella should consist of 4 to 5 layers of extended chain macromolecules. The isotropization temperature of the mesophase lamellae must be dependent on the isotropization temperature of a single layer consisting of extended macromolecules, i.e. it must depend on their length. Therefore, a simplified Gibbs-Tompson equation for the melting point of lamellar crystals can

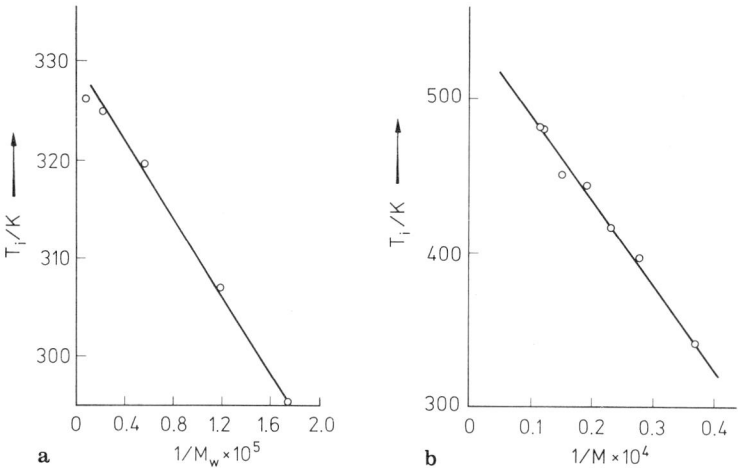

Fig. 14a and b. Dependence of the isotropization temperature T_i on the reciprocal molecular weight [22,27]. **a** PDES, **b** PDPS

be used for an estimation of the isotropization temperature T_i. This equation has a form

$$T_i = T_i^0(1 - 2\sigma_e/(l\Delta H_i)) \qquad (4)$$

where T_i is the isotropization temperature of the lamellar mesophase crystallite, T_i^0 is the isotropization temperature of the mesophase crystallites of the infinite thickness, l is the lamellae thickness (more correctly the thickness of the mesophase layer of the lamella) and ΔH_i is the isotropization enthalpy. Such a dependence has really been observed for both PDES and PDPS (Fig. 14). The values of Δ_e are 20 erg/cm² for PDES and 310 erg/cm² for PDPS. While the value of Δ_e for PDES is quite reasonable, the σ_e for PDPS is unreasonably high. Although this consideration shows that the dependence of T_i on molecular mass for the linear polysiloxanes can be explained, in principle, because of a very low enthalpy of isotropization and rather high values of the interface energy of the mesophase layers of extended chains forming the mesophase lamella, for a final conclusion further investigations on more narrow fractions especially in the region of low molecular masses are required.

2.1.6 Effect of the Mesophase on the Formation of the Crystalline State

One of the very interesting problems of the mesomorphic state is the effect of a preliminary ordering resulting from the mesophase formation on crystallization. The problem was studied on PDES with DSC and the picture revealed is as follows (Fig. 15) [21]. When the cooling rate of the isotropic melt is so high that the mesophase can not be formed during cooling, the crystallization is similar to the process which occurs in polymers with a moderate crystallization rate. Indeed, upon such a quenching, the melt of PDES can be transformed into the amorphous glassy state. In this case, on the subsequent heating of the quenched sample a jump

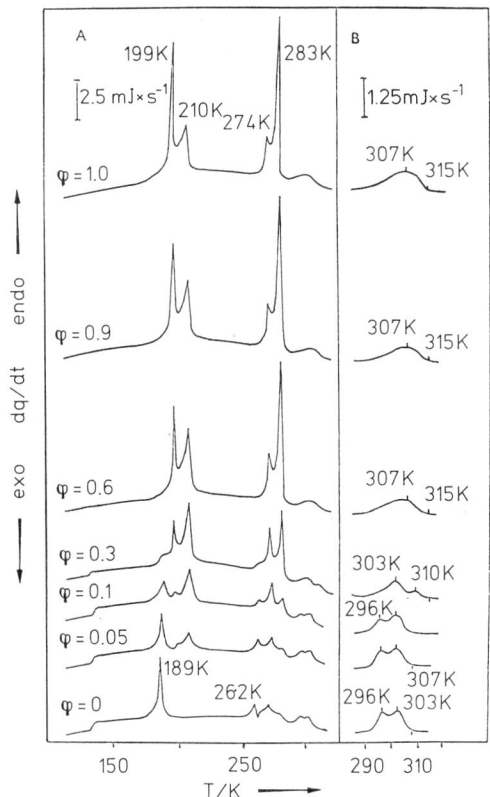

Fig. 15. The (A) complete DSC records and (B) enlarged mesophase isotropization peak of PDES samples crystallized at different degrees of conversion to the mesophase φ by cooling at the rate of 320 °C/min. Heating rate, 10 °C/min [27]

in the heat capacity ot T_g and a cold crystallization occurs. The degree of crystallinity which can be reached in this case is typical of crystalline polymers (30 to 40%). Quite a different picture appears when the formation of the mesophase occurs before crystallization. In this case, the sample cannot be transformed into the amorphous glassy state even by quenching because of an extremely high crystallization rate of the mesophase. Typical values of the degree of crystallinity reached exceed 90%. Remembering that linear polysiloxanes are typical elastomers the effect of the mesophase on the crystallization is quite obvious.

In contrast to PDES, it is impossible to obtain PDPS in the completely amorphous glassy state even by quenching its melt with liquid nitrogen. For PDPS, due to a large interval between the temperature of the mesophase formation from the melt and crystallization (see Fig. 4); at any cooling rates, the mesophase is already formed before crystallization occurs. Hence, the presence of the mesophase in the temperature interval of crystallization leads to such fast crystallization rates that the possibility to reach the amorphous state or even obtain the mesomorphic glassy state is excluded. In PDES, amorphization seems to be possible because of a small temperature range of existence for the mesophase. The behaviour of PDPS also supports the fact that the formation of the mesophase from the isotropic melt is controlled by a process with a very large negative temperature coefficient. Such a process is undoubtedly the formation and growth of the mesophase nuclei.

A well-defined lamellar morphology of the mesophase state and the effect of the mesophase on the crystallization entails, in turn, the problem of structural reorganization occuring during crystallization. As it has been mentioned above, one of the crystalline forms, namely α_2, bears a close relationship to the mesophase structure and the phase transition does not require considerable reorganization. These findings agree with the results of the optical observations [35,37], according to which, crystallization occurs as a development of birefringent mesophase domains. It means that the initial texture of the mesophase samples largely remains intact, i.e. the spatial arrangement and orientation of the lamellar domains is retained. Crystallization occurs following a certain induction period. Initially, crystallization centers appear and then they start to grow, their number being constant throughout the crystallization process.

The growth of the crystalline regions in different directions is not uniform, resulting in their largely irregular geometry. The growth is not steady in time: some of the regions cease to grow for a certain period, then they resume their growth, as a rule, upon coming into contact with other growing crystalline regions. Like the crystallization of polymers from the amorphous state, the crystallization of PDES proceeds via the formation of crystalline nuclei and their subsequent growth. Indeed, the heat evolution kinetics at 276 K may be described by the Avrami equation with the morphological parameter $n = 2.1$ (Fig. 11). This may be interpreted as the lamellar growth of crystallites on athermal nuclei. Unfortunately, due to the strong dependence of crystallization kinetics on the conditions of the mesophase formation and on the time and temperature of the mesomorphic sample stand prior to crystallization, the crystallization energy parameters could not be determined. Similar difficulties occured during the study of the crystallization of the mesophase in PDPS [38].

In conclusion it has to be emphasized that crystallization from the mesomorphic state occurs at low supercooling temperatures, not as the sporadic crystallization of individual mesomorphic lamellae, but as the growth of the crystalline regions via the consecutive incorporation of the crystallizing lamellae. Crystallization of each lamella that comes into contact with the crystalline region seems to occur only when it is induced by the strain arising at the phase boundary or is the result of epitaxy in the case of coplanarity of the lamellar faces and the crystallization front. Otherwise, the contacting mesomorphic lamellae may fail to crystallize and the advance of the crystallization front in the same direction may stop. This is likely to be the reason for a strong effect of the thermal history of the sample on the transition kinetics.

2.1.7 Elasticity and Thermoelasticity

The inclination of PDES macromolecules to form the mesophase leads to a peculiar stress-strain behaviour of crosslinked samples above T_m. The stress-strain curves of semi-mesomorphic crosslinked PDES films resemble the stress-strain curves of the crystalline polymers with a low degree of crystallinity which deform without necking. Figure 16 shows typical curves for slightly crosslinked films at 293 K [22,39]. The same film which was preliminarily amorphosized at 353 K behaves in a completely different manner. At relatively small elongations ($\lambda = 2.3$ to 3 depending on tem-

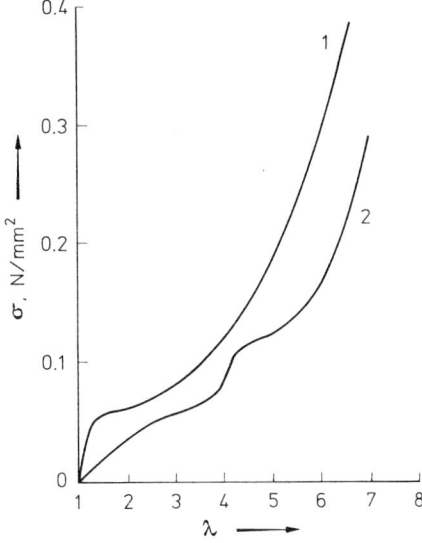

Fig. 16. Stress-strain curves of crosslinked PDES film [39]. Room temperature. 1 — mesomorphic film, 2 — completely amorphous film. Stretching rate — 0.065 mm/s

perature), the film is elongated as a typical non crystallizable network. At this initial stage, the stress relaxation and hysteresis effects are absent and the curve is insensitive to the rate of the deformation. Moreover, this part of the stress-strain curve follows the Money-Rivlin equation very well. A rather unusual feature of the deformation behaviour consists of a very large increase of the stress with temperature. In fact, the stress at 327 K, that is at the final stage of the isotropization, is almost two times higher than at 293 K. Analysis of the values of C_1 and C_2 in the Mooney-Rivlin equation showed that it is a consequence of an unusually large decrease of C_2.

A further extension induces the formation of the mesophase. Its appearance leads to a constant or even decreasing stress. Finally, a sharp increase of the stress occurs on the curve. In general, the shape of the stress-strain curve is determined by the relation of the deformation rate and the rate of the formation of the mesophase. At lower temperatures, the formation of the mesophase is a rather fast process and the mesomorphic regions which appeared at $\lambda > 2$ seem to be oriented at some angle to the stretching direction. A further extension is accompanied not only with the appearing new mesophase regions, but also with reorganization of these already existing regions. This process needs some additional mechanical work and leads to the three stage stress-strain curve. The third part of the curve corresponds to the extension of the oriented mesophase sample. At increased temperatures the mesophase occurs at the elongations when chains are well extended (oriented) and the mesophase domains are oriented preferentially along the stretching direction. At such temperatures, the S-shaped part is absent and the two-stage stress-strain curve looks like that of the crystallizable networks. Similar behaviour was also established for the PDES samples prepared by means of multifunctional crosslinking agents [40].

An additional important piece of information concerning the deformational behaviour of the mesomorphic PDES networks was obtained with thermoelastic (thermomechanic) measurements [22,39,40]. The direct calorimetric measurements of

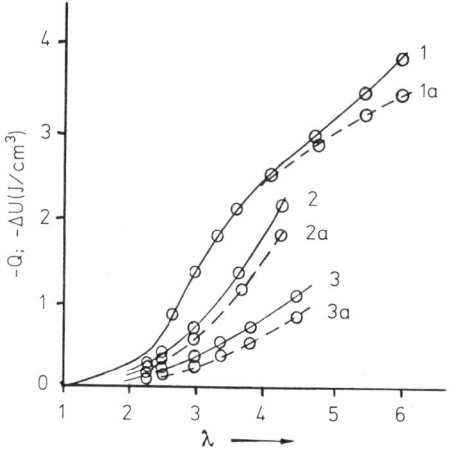

Fig. 17. Heat (1, 2, 3) and internal energy change (1a, 2a, 3a) on stretching PDES films from the unstrained state to λ at different temperature [39]. 1, 1a and 3, 3a — film with cross-link density 3.2×10^{-6} mole/cm^3; 2, 2a — film with cross-link density 6.5×10^{-6} mole/cm^3. 1, 1a — 293 K; 2, 2a and 3, 3a — 327 K

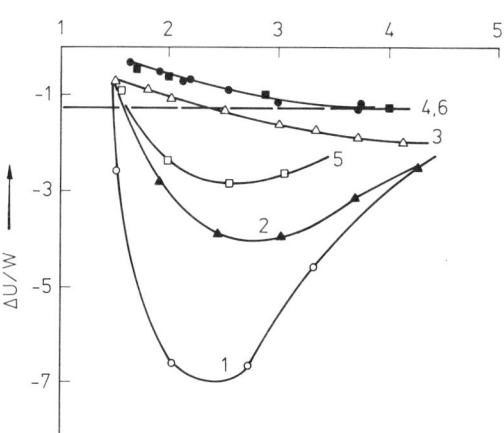

Fig. 18. The calorimetrically determined relative internal energy contributions as a function of elongation ε [40]. 1, 2, 3, 4 — network with 15% of a multifunctional crosslinking resin; 5, 6 — network with 45% of a multifunctional resin. Temperatures: 1 — 20°, 2 — 30°, 3 — 50°, 4 — 70° and 100 °C; 5 — 20°, 6 — 80 °C. The dotted line represents intramolecular (conformational) energy contribution for PDES

the heat and energy effects resulting from the stretching of the PDES networks by deformational calorimetry [41] showed that the characteristic feature of their thermomechanical behaviour is a large decrease of the internal energy ΔU. Typical thermomechanical curves are shown in Figs. 17, 18. Even at the initial extensions (λ < 2) when the formation of the mesophase is excluded (according to the stress-strain curves), the internal energy to work ratio is close to −1.2. It is an unusually high value for the energy contribution in comparison to other networks studied [41,42] corresponding to a very high negative temperature coefficient of unperturbed chain dimensions $d(\ln \langle r^2 \rangle_0)/dT = -4 \times 10^{-3}/\deg^{-1}$.

It is noteworthy that the sign of the energy contribution in PDES is negative unlike PDMS chains in which f_u/f or $(\Delta U/W)_{V,T}$ is $+0.28$ [41,42]. The negative value of the energy contribution in PDES means that the extended conformations of macromolecules are energetically more advantageous. One can suggest that this is the reason of the ability of PDES macromolecules to form the mesophase.

At 293 K, the absolute value of the $\Delta U/W$ ratio at the initial stage of the stretching curve for the slightly crosslinked networks is even higher and close to -8. It is interesting to note that a considerable heat relaxation takes place in this case, although the deformation is completely reversible and the stress relaxation is absent. This thermal relaxation seems can be partly connected with the volume relaxation.

Using the integral change of the internal energy (Fig. 16) the increase of the mesophase content during stretching can be estimated. It is seen that in the slightly crosslinked films the content of the mesophase can approach 0.7 while in the initial unstretched samples the content of the mesophase at room temperature is not higher than 0.3 to 0.4. Hence, stretching promotes the formation of the mesophase and an obvious analogy exists between the mesophase formation and the crystallization of crystallizable networks under stretching. Further thermoelastic and thermomechanical studies are necessary to establish whether the theories developed for crystallizing networks can be successfully used for thermodynamical treatment of formation and melting of the mesophase in the stretched mesomorphic networks.

2.1.8 Phase Behaviour of the Mixtures of Polydiethylsiloxane with Oligomers of Dimethylsiloxanes

Miscibility experiments in binary mixtures are an important method of examining and classifying the liquid crystal phases in low molecular substances and polymers [6]. In the only similar study of the mesophase behaviour in PDES at present the phase diagrams of PDES with oligomers of dimethylsiloxanes were obtained [43]. Figure 19 represents that part of one of the diagrams which demonstrates the mesophase behaviour. Besides the transitions found in PDES in the mixtures of PDES and oligodimethylsiloxanes additional transitions occur. As can be seen from the diagram;

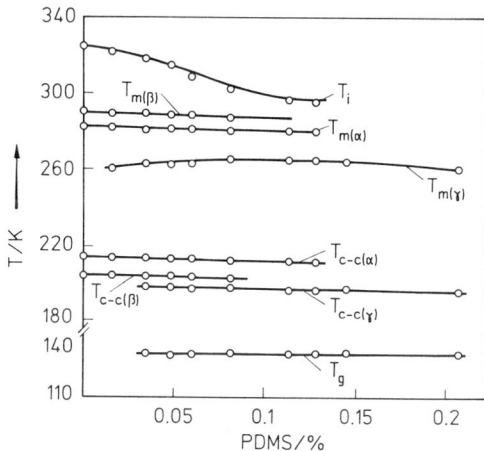

Fig. 19. Dependence of the glass transition temperature T_g, solid-solid transition temperatures T_{c-c} for corresponding crystal forms, melting points T_m for corresponding crystal forms and isotropization temperature T on the content of PDMS oligomer in blends [43]

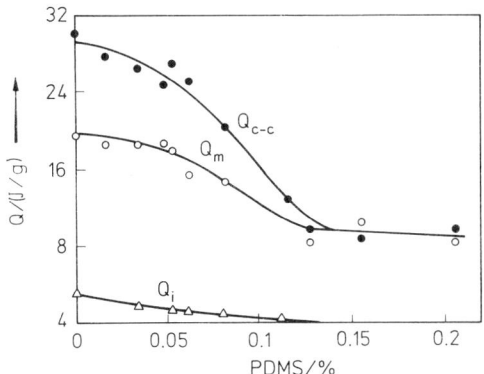

Fig. 20. Dependence of the heat of solid-solid transition Q_{c-c}, heat of fusion Q_m and heat of isotropization Q_i on the content of PDMS oligomer in blends [43]

with the increase of the oligomer content, T_i decreases considerably; and above approximately 12% of oligomer the mesophase disappears completely. Simultaneously the two high-temperature and two lowtemperature transitions vanish. This seems to indicate that a close relation exists between the crystalline phases and the mesophase. This conclusion is in full accord with the one concerning the X-ray crystalline structure and the structure of the mesophase. With increased content of the oligomer in the mixture, the all the heats of transitions decrease considerably as can be seen from Fig. 20. Above the critical content of the oligomer the heats of the crystalline transitions reach practically constant values. Estimation of the Flory-Haggins interaction parameter showed that $\varkappa_{12} = 0$ which can be expected for the components with a similar chemical structure.

2.2 Cyclolinear Polyorganosiloxanes

2.2.1 General Description

One of the striking recent achievements of organosilicone chemistry is the synthesis of cyclolinear polyorganosiloxanes (CLPOS). The synthesis of the polymers which include two or multistep heterofunctional polycondensation reactions is described in detail [44–46], therefore, we will not touch the chemical aspects of the polymers in this article. CLPOS consist of linear chains of repeated siloxane cycles of various dimensions bonded by oxygen atoms. The general formula of this class of polymers is

$$\left[\begin{array}{c} O-[Si(R)_2O]_n R' \\ -Si Si-O- \\ R' O-[Si(R)_2O]_m \end{array} \right]_p$$

$R = CH_3, C_2H_5, C_6H_5$; $R' = CH_3, C_2H_5, C_6H_5$; n, m = 1, 2, 3. The number of the repeat cycles in macromolecules may be from some dozens to several hundreds. The substituents R in the organosilsesquioxane fragment can possess various spatial arrangements which lead to the polymers of different stereotacticity, namely the cis-, trans- and atactic polymers. The study of the hydrodynamic properties of some CLPOS

Table 4. Dependence of Khun Segment on the chemical structure of CLPOS

Polymer	Khun segment Å
[cyclotetrasiloxane with Me substituents]	29
[cyclohexasiloxane with Me substituents]	23
[larger cyclosiloxane with Me substituents]	40
[cyclosiloxane with Me and Ph substituents]	110

gives the possibility of estimating their chain flexibility [47]. As one can see from Table 4, the polymers with methyl substituents have the Kuhn segment typical of the flexible polymers. Such macromolecules are really rather flexible. The incorporation of the phenyl substituents leads to a considerable increase of the Kuhn segments which can reach the values typical of some ladder polyphenylsiloxanes [11]. The latter polymers are known to be characterized by a considerably higher rigidity in comparison with the typical flexible polymers.

Recent investigations of some properties and structure of CLPOS showed that the polymers with the cycle dimension $n = m = 1$ (cyclotetrasiloxanes) and $n = m = 2$ (cyclohexasiloxanes) are able to form thermotropic mesophases [44,48-52]. Hence, CLPOS is a new class of the mesomorphic polymers. The interest in this class of the mesomorphic polymers is connected, first of all, with the possibility of following the characteristic changes in the mesomorphic state resulting from the increase in the rigidity of the macromolecules which as can be seen from Table 4 can change considerably. On the other hand, the comparison of their mesophase behaviour with the mesomorphic state of linear polyorganosiloxanes must throw some light on the nature of their mesomorphic state. Finally, CLPOS are of great interest as a new class of the mesomorphic polymers.

2.2.2 Phases and Phase Transitions

The first evidence of the mesophase behaviour of CLPOS was obtained with scanning calorimetry and X-raying of the stereoregular syndiotactic cyclolinear methyltetra- and methylhexasiloxanes [44,48]. A typical thermogram and diffractogram is shown in Fig. 21. Besides glass transition (T_g = 182 K) and melting (T_m = 220 K) in the sample slowly which was cooled from room temperature to 150 K, there were small well-reproduced endothermal peaks with maximums at 250 and 380 K and a rather large endothermal peak with the maximum at 595 K. X-ray patterns show that below 220 K the polymer is a typical semicrystalline polymer with a low degree of crystallinity (approximately 10%) estimated from the heat of fusion. In the temperature range 220 to 595 K, the X-ray pattern is characterized by only one sharp reflection in the range 2θ = 8 to 12ë and the amorphous halo corresponds to a typical mesomorphic structure. Above 595 K, the diffraction profile spectrum of this polymer is similar to that of amorphous polysiloxanes. The X-ray picture of the

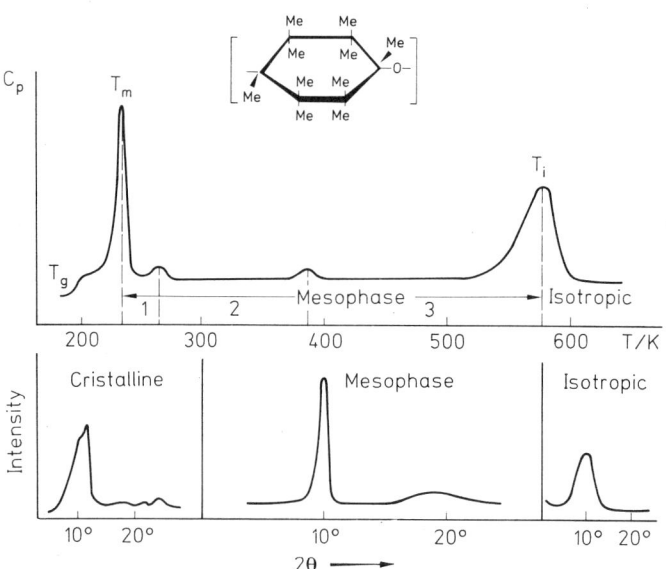

Fig. 21. DSC-curves of slowly cooled stereoregular syndiotactic poly[oxy(decamethylcyclohexasiloxane-2,8-diyl)] and X-ray spectra in crystalline, mesophase and isotropic states [48–51]. Heating rate is 10 K/min

mesomorphic and isotropic state in this CLPOS is qualitatively analogous to that of linear polysiloxanes (see Figs. 2, 5). In the polarizing microscope under cross polarizers, characteristic birefringent areas of typical size 5 to 50μ were observed in the temperature interval from room temperature to 600 K. During heating, a distinct change of birefringence at 370 to 390 K has been observed and above 500 K birefringence gradually disappears. Hence, according to calorimetric, X-ray and optical observations the transition at 595 K may be identified as the isotropization temperature. The transitions at 250 K and at about 380 K were suggested to correspond to the transformation of one type mesophase (not yet identified) to another. Although any visible changes of the sharp reflection are absent at these temperature transitions, a clear change in the positions of the amorphous halo under this reflection was observed.

Finally, it is necessary to note that like linear PDPS this polymer cannot be transformed into the amorphous state by quenching of the isotropic melt. Thus, in spite of the difficulties of identification of the types of the mesophases and the mesophase transitions at heating, one can only definitely conclude that the temperature interval of the mesophase stability of this polymer is greater than 350 K. The temperature interval of the mesophase is strongly dependent on the cycle dimension. In fact, in cyclolinear polytetramethylsiloxane the temperature interval of the mesophase stability is only about 10 K [44].

2.2.3 Effect of Substituents

The existence of the mesomorphic state is characteristic not only of the methyl CLPOS but also of the polymers with other substituents [51, 53, 54]. The effect of phenyl substituents on the thermal behaviour of a series of phenyl CLPOS is shown in Fig. 22. The introduction of phenyl groups into the siloxane cycle resulted in an increase of both the melting point and the glass transition temperature. An interesting conclusion can be drawn as a result of the comparison of the cyclolinear polymers and the linear ones (cf. Fig. 22). The cyclolinear polymer with all methyl

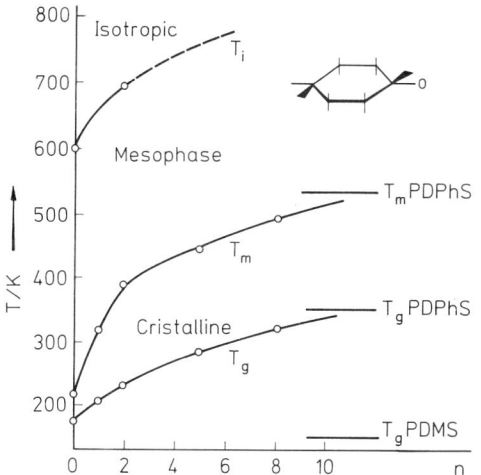

Fig. 22. Dependence of the glass transition temperature T_g, melting point T_m and isotropization temperature T_i on the number of phenyl groups n in cycles of cyclolinear polymer [53, 54]. Horizontal solid lines indicate the corresponding transitions for PDPhS and PDMS

substituents has a T_g about 30 K higher than the linear PDMS. This seems to indicate the restriction of the local molecular mobility arising from the joining of the six linear Si—O bonds into the cycle, viz., the existence in the chain of the organosilsesquiooxane fragment. On the other hand, on the complete substitution of the methyl groups by phenyl ones the T_g of the polymer is about 20 K below the T_g of linear PDPhS. In terms of the free volume concept, it means that the packing of the cyclolinear chains with phenyl groups is less dense than that of the linear chains. All the cyclolinear polymers with phenyl substituents are able to crystallize. The introduction of the first and second phenyl substituents is accompanied by a very large increase of the T_m. The T_m of the fully phenyl polymer is only 30 K below the T_m of PDPhS. The heat of fusion increases monotonically from 5.5 J/g for the methyl polymer to 33.5 J/g for the phenyl polymer, which is rather close to the heat of fusion of PDPhS.

Above the T_m, all the polymers shown in Fig. 22 can exist in the mesomorphic state. This conclusion follows, first of all, from the X-ray patterns which possess the single narrow intensive reflection with $2\theta = 8$ to $12°$ typical of the mesomorphic state of the siloxane polymers. The temperature interval of the mesophase stability is strongly dependent on the number of phenyl groups introduced into the cycle. The T_i of the polymers with a large number of phenyl groups seems to be above their intensive chemical degradation. Unlike the heat of fusion, the dependence of the heat of isotropization on the number of phenyl groups is of a nonmonotonic character. The introduction of the first phenyl substituent is accompanied by a sharp decrease of Q_i from 10 to 3.8 J/g after which a very slow decrease of Q_i occurs. Hence, the systematic substitution of methyl groups by phenyls in CLPOS does not lead to a lowering of the ability to form the mesomorphic state. The change of the ratio methyl to phenyl groups in cycles leads to a change of the temperature range of the mesophase stability and seems to influence the degree of the mesomorphic state (degree of mesophasity).

The similar dependence of the transition temperature on the number of phenyl substituents in the cycle is also found for polymers with tetrasiloxane cycles although the temperature interval of the mesomorphic state in this case is considerably narrow [54, 55].

2.2.4 Effect of the Chain Tacticity and Flexible Spacers

It is well-known that chain molecules are generally able to crystallize only if they are stereoregular (syndiotactic or isotactic). On the other hand, polymers with an irregular alternation of monomeric units with various stereometric configurations along the chain are not capable of crystallizing. One can suggest that a strict chemical regularity of the macromolecules is not a necessary condition for its appearance. The effect of the various chain tacticity on the mesophase behaviour was clarified by a systematic study of transtactic, isotactic and atactic cyclolinear decaorganocyclohexasiloxane polymers with methyl and phenyl substituents of the silicon atom in the organosilsesquixane fragment. The following picture was revealed as a result of the investigation of the effect of the chain tactivity [46]. It has been established that: (i) T_g is dependent only on the chemical composition and independent of the spatial configuration of macromolecules; (ii) transtactic polymers crystallize (even under

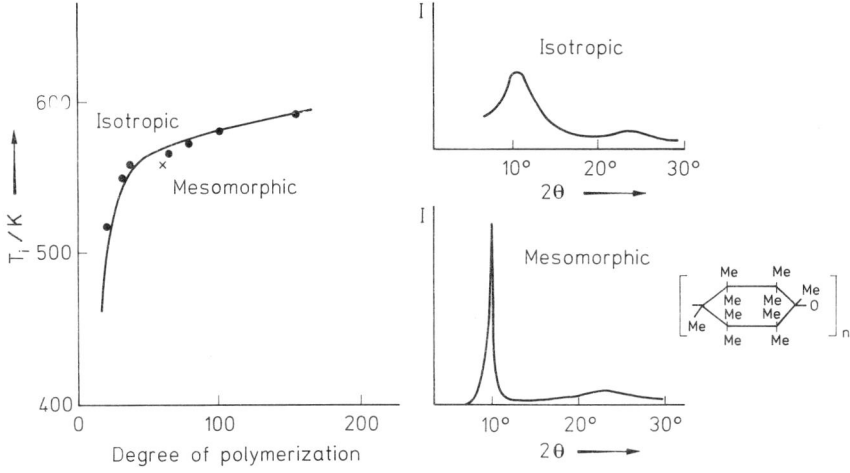

Fig. 23. Plot of the isotropization temperature T_i as a function of the degree of polymerization of atactic poly[oxy(decamethylcyclohexasiloxane-2,8-diyl)] and X-ray spectra in mesomorphic and isotropic states [52]

fast cooling) but cistactic and atactic polymers do not crystallize; (iii) the mesomorphic state exists irrespective of the chain stereotacticity, T_i being dependent on microtacticity. Although the values of T_i of the polymers with various tacticity differ considerably, it is difficult to elucidate the role of the polymer tacticity on T_i immediately, because of the simultaneous strong influence of the molecular mass on T_i as one can see from Fig. 23 where a typical dependence is shown. Taking into account the T_i dependence on molecular mass, it was established that at a comparative molecular mass the highest clearing point (isotropization temperature) is characteristic of transtactic polymers, the lowest — of cistactic, and atactic polymers possess intermediate values of T_i. The rate of the mesophase formation under cooling of the isotropic melt is rather high, therefore, cistactic and atactic polymers below T_g are the mesomorphic glasses. Returning to the dependence of T_i in atactic polymers on the molecular mass it is necessary to note that the intensive increase of T_i for the cyclolinear polymer takes place approximately in the same range of molecular masses as it does for PDPS, although their absolute values of T_i are considerably higher. The analysis according to Eq. (4) showed that the slope of the line is close to that of PDPS which, in turn, leads to extremely high values of σ_e. As for PDPS, the reason for such high values of σ_e is not clear yet.

Unlike poly-decaorgano-cyclohexasiloxanes which form the mesomorphic state independent of the chain tacticity, poly-octaorgano-tetrasiloxanes can exist in the mesomorphic state only in transtactic configuration.

These facts seem to emphasize an important role of the hexasiloxane cycle irrespective of its tacticity and the nature of the substituents in forming the mesophase state and enable even the consideration of this cycle as an original "mesogen". The peculiarity of the "mesogen" consists in that, unlike the classical mesogens, it is capable of forming a mesomorphic phase of attainable T_i, being introduced into the main chain without any flexible spacer. This seems to indicate that this cycle is relatively flexible, therefore, the introduction of an additional flexible spacer between adjacent cycles can considerably change the mesophase stability, above all, because of the depression of T_i. Moreover, with relatively short spacers, the mesomorphic phase can disappear completely. The checking of the validity of this suggestion was made on transtactic cyclohexasiloxane polymers in which between the every cycle dimethylsiloxane spacers of various length were incorporated [54, 56]. The general formula of such polymers is

$$\left\{ \begin{array}{c} R-Si \underset{O-[(CH_3)_2 SiO]_2}{\overset{O-[(CH_3)_2 SiO]_2}{\diagup}} Si \underset{O-[(CH_3)_2SiO]_n-}{\overset{R}{\diagup}} \end{array} \right\}_p$$

$n = 0\text{--}4$, $R = CH_3$, C_6H_5.

These polymers may evidently be also considered as linear PDMS polymers with the cycle fragments in the main chain. The results listed in Table 5 support the suggestion concerning the decisive role of the flexible spacers not only on the stability of the mesophase state but also on the ability to crystallize. These results seem to show that the ability of the hexasiloxane cycle with methyl substituents to form the mesomorphic state is connected not with its rigidity, as in classical mesogens, but with some packing factors. As far as the cycles with phenyl substituents are concerned both the enhanced rigidity of the cycles and packing factors seem to play an important role.

Table 5. Thermal data for methylsiloxane polymers with decaorganodioxycyclohexasiloxane fragments [56]

R	n	$[\eta]$ dl/g toluene	T_g, K	T_m, K	ΔH_m, J/g	T_i, K	ΔH_i, J/g
CH_3	0	0.15	182	220	5.5	595	10.0
	1	0.13	185	285	17.6	370	3.6
	2	0.14	182	—	—	330	1.3
	3	0.15	183	—	—	—	—
C_6H_5	0	0.21	232	460	—	700	—
	1	0.19	224	355, 380	16.8	440	1.3
	2	0.15	206	325	14.7	—	—
	3	0.14	196	—	—	—	—

2.3 Ladder Polyorganosiloxanes

The extreme case of a cyclolinear siloxane chain is ladder polyorganosylsesquioxane macromolecules (POSSO)

$$\begin{bmatrix} \overset{R}{\underset{|}{|}} & \overset{R}{\underset{|}{|}} & \overset{R}{\underset{|}{|}} \\ -Si-O-Si-O-Si- \\ | & | & | \\ O & O & O \\ | & | & | \\ -Si-O-Si-O-Si- \\ | & | & | \\ R & R & R \end{bmatrix}_p$$

At present, various POSSO with different aryl substituents R_1 on silicon atom and even containing in some proportion alkyl side groups R_2 have been synthesized [57,58]. Cis-syndiotactic conformation of the double chain macromolecules POSSO is assumed [57,59]. According to viscosity and dynamooptic measurements, the length of the Kuhn segment of POSSO macromolecules varies from 100 to 200 Å [59]. The macromolecule diameter is equal to about 10 to 12 Å. Consequently, the axial ratio characterizing the asymmetry of semirigid chains of POSSO is 10:20 and, as predicted by theory [4], such macromolecules even in the absence of specific interaction of side groups should form liquid crystalline phases. An analysis of the data published leads indeed to the conclusion that POSSO similar to cellulose derivatives, may be considered as an example of lyotropic mesophases formed by semiflexible macromolecules [60]. Unfortunately, precise data on the separation of the mesophase from POSSO solution are not available. This is connected in part with the fact that, due to the relatively low values of the axial ratio, the formation of an ordered phase in solution proceeds at high concentrations of POSSO. For example, in poly(m-chlorophenylsylsesquioxane)-benzene system, light opalescence appears at the volume fraction of the polymer not less than 0.9 and well-defined anisotropic domains are formed only after a total removal of the solvent. Therefore, the main attention was paid to the investigation of the morphology of mesomorphic films of POSSO casted from solutions [61-63].

Due to high viscosity of any concentrated solutions the separation of the ordered phases is kinetically hindered, therefore, the ordering in such films can proceed to a different extent depending on casting conditions. In particular, the morphology is appreciably affected by the solvent type. From the same polymer using different solvents, it is possible to produce both transparent isotropic films without visible phase separation and turbid films with ordered optically anisotropic domains.

Similar to other mesomorphic polymers, diffractograms of POSSO consist of a sharp reflection characterizing the intermolecular spacing and an amorphous halo (Fig. 24). The intensity and sharpness of the first reflection depend on the conditions of sample preparation. According to the light scattering, the ordered domains have an asymmetric shape. Under the microscope between crossed polars one can observe optically anisotropic regions of some microns in size. Electron micrograph replicas of the film suface show that the morphology of POSSO depends on their molecular mass which is clearly seen in Fig. 25. The lower molecular mass fraction forms well defined anisotropic supermolecular structures consisting of fibrous subunits aligned parallel to the longitudinal axis of these structures. The mesomorphic domains arising in a more viscous solution of higher molecular mass fraction have another shape, namely, they look like bent bundles.

The presence of the mesophase domains in films of POSSO markedly influences

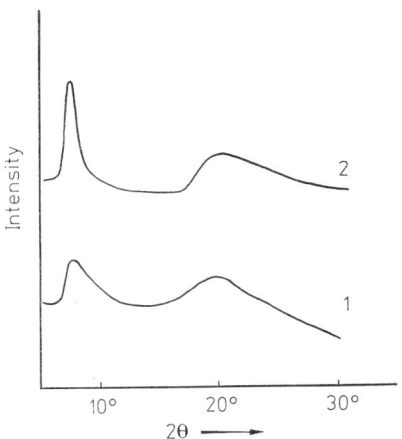

Fig. 24. Diffractograms of a transparent optically isotropic film (1) and a turbid film with optically anisotropic domains (2) of polyphenylisobutylsilsesquioxane [62]. The curves are shifted arbitrarily

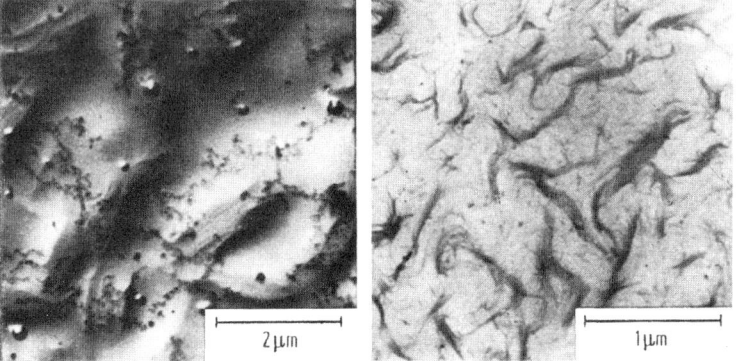

Fig. 25a and b. Electron micrographs of the free surface of films casted from a benzene solution of two fractions of polyphenylisobutylsilsesquioxane with the molecular weight 100,000 **a** and 200,000 **b** [63]

their mechanical behaviour and thermostability [64, 65]. On heating, the mesomorphic domains disappear in the temperature region of 620 to 670 K and, simultaneously, polymer degradation starts. Because of this, it is impossible to minitor the thermotropic transition into the isotropic state by scanning calorimetry. However, the difference in the internal energy between the isotropic and mesomorphic state of POSSO is 10 to 15 J/g, as it follows from the heats of dissolution [66].

The formation of the mesophase in POSSO can also occur during their synthesis. This process was investigated in detail during polymerization of the cage-like m-tolylsilsesquioxane cycles into poly(m-tolylsilsesquioxane) macromolecules [67, 68]. The polymerization process proceeds as a consecutive addition of cage-like cycles $[CH_3C_6H_5SiO_{1.5}]_n$ (n = 6 to 12, $\bar{M}_n$ = 2,000) to the active sites at the ends of living macromolecules via anionic mechanism. Electron micrographs of replicas of the fracture surface of bulk sample taken at different stages of polymerization are shown in Fig. 26. Starting with the certain stage when the transparent polymerizing

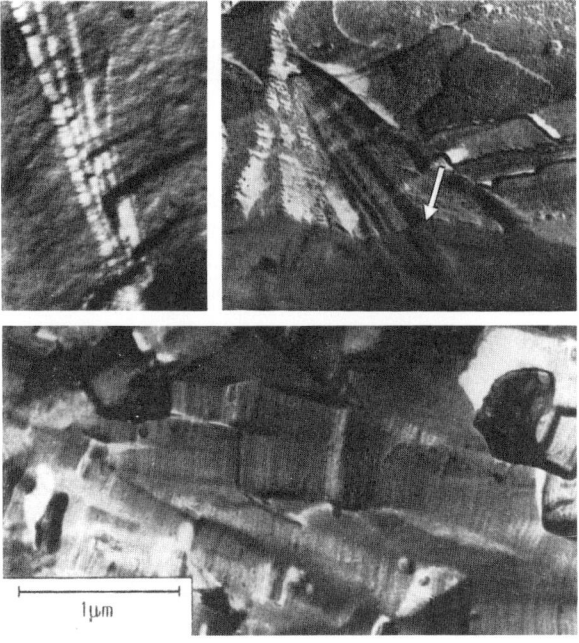

Fig. 26 a–c. Electron micrographs of a fracture surface of a bulk sample of poly-*m*-tolysilsesquioxane taken at different stages of polymerization [67]. **a** polymerization time 3 min at 220 °C; **b** polymerization time 9 min at 220 °C; the free surface (bottom) and fracture surface (top) are shown, the arrow indicates the fracture edge; **c** polymerization time 30 min at 250 °C

melt becomes turbid, lamellae of striated appearance can be recognized on the fracture surface. These lamellae are seen more clearly if they are arranged perpendicularly to the fracture surface. The transversely striated lamellae consist of growing macromolecules arranged perpendicular to the end surface of the lamellae. An increase of lamellae thickness during polymerization confirms this suggestion. In fact, in 6 min. of polymerization at 493 K the lamellae thickness reaches 600 to 700 Å and in 9 min. 700 to 1000 Å (Fig. 26). These thicknesses correspond to molecular masses of 70,000–80,000 and 80,000 to 110,000, respectively. The GPS data were in good agreement with the above values. The lamellae get large longitudinal dimensions. Such lamellar structure is especially recognizable in Fig. 26, where an end edge of a fractured sample with a free and a fracture surface is shown. End surfaces of lamellae emerged on the free surface give a specific picture. At the final stage of polymerization, lamellae resemble the striated lamellae observed in some crystalline polymers. According to the electron micrographs, the polymer at this stage is an agglomerate of randomly spaced and grown-together lamellae. The thickness of many single lamellae is 1300 to 2000 Å which corresponds to molecular masses 150,000 to 200,000. It coincides with the molecular mass 180,000 evaluated from GPC. Typical mesomorphic extended chain lamellae form only during polymerization. Morpholgy of the solution casted films of the polymer is similar to that of other solution casted POSSO.

According to the shape of supermolecular ordered structures observed on the surface of POSSO films, these mesomorphic systems are apparently close to the nematic liquid crystals. On the other hand, the above extended chain mesomorphic lamellae arising during polymerization may be properly regarded as a macromolecular analog of smectic liquid crystal. Although the increased rigidity of POSSO chains seems to be a sufficient factor for forming the mesophase, the specific interaction of the side aryl and alkyl groups can undoubtedly favor this process which probably takes place in the case of flexible linear and cyclolinear siloxane polymers. Further, more detailed studies of the structure of mesophases in cyclolinear siloxanes must elucidate a specific role of the chain rigidity in forming POSSO mesophases.

2.4 Low Molecular Weight Siloxanes

For better understanding of the nature of the mesophases in element-organic polymers, it is very useful to consider the question of what is currently about the mesomorphic state of low molecular siloxanes. An analysis of the not numerous relevant publications showed that there are some good examples of the thermotropic mesophases in substances of this class. Moreover, there is some similarity between the mesophase state of low molecular siloxanes and polyorganosiloxanes, viz., these mesophases differ from the classical nematic and smectic liquid crystals, they seem to resemble closely the disordered crystals, but their exact classification is difficult. Decaethylcyclopentasiloxane [69] and diisobutylsilandiol [70] are apparently the first examples of thermotropic mesomorphic siloxanes. It has been found that the former transformed from the crystalline state into the isotropic melt via an intermediate ordered state. It has been suggested that in this intermediate phase the orientational order disappeared and the molecules started rotating freely as in the plastic crystals. The latter siloxane is also semi-solid in the mesophase state. Its structure was studied in more detail [71, 72]. According to the data from DSC the heat of transition from the crystalline to mesophase state at 88.4 °C was 7.6 kJ/mole and the heat of transition from the mesophase to the isotropic melt at 98.7 °C was 7.2 kJ/mole. Eaborn and Hartshorn [71] found that although it showed fan-like and striated-band optical texture which is normally characteristic of smectic phases, the mesophase was optically negative. There was, however, no sign of the characteristic optical properties of cholesteric phases. The formation of the mesophase by i-$Bu_2Si(OH)_2$ was tentatively interpreted in terms of hydrogen-bonded chains of molecules, involving hydrogen bonds of bifurcated type thought at that time to be present in diethyl- and diallylsilane diols. It was suggested that the mesophase consists of parallel layers which could deform and glide over one another as in ordinary smectics but with negative anisotropy. However, later X-ray study [72] has shown that the optical texture of the mesophase in i-$Bu_2Si(OH)_2$ may have another origin. The X-ray diffraction pattern of the mesophase is rather simple. It contains an outer diffuse ring corresponding to a repeat distance 4.7 Å and an inner ring corresponding to a repeat distance of approx. 11 to 12 Å. The inner ring is of an intermediate type. It is by no means as sharp as those found for smectic mesophases, but it is not as diffuse as the ring observed for isotropic samples for nematic mesophases. The only type of mesophase which has been observed to give a diffraction pattern of this type is the discotic phase. It was also found that the optical texture of the mesophase is similar to a great extent to those of the discotic phases of benzene hexa-n-alkanoates. From these findings and the results of detailed studies of the crystalline structure of di-t-butylsilandiol it was suggested that the mesophase of i-Bu$_2$Si(OH)$_2$ is of the discotic type in which the six-number ring of the dimer serves as a discotic unit. The thickness of this unit is about 5 Å (the repeat distance along Si—H ... HO—Si chains) and the diameter of the dimer in the plane in which butyl side groups lie is about 12 Å. In the crystalline state, the dimers are connected by hydrogen bonds forming ladder chains as it is shown below.

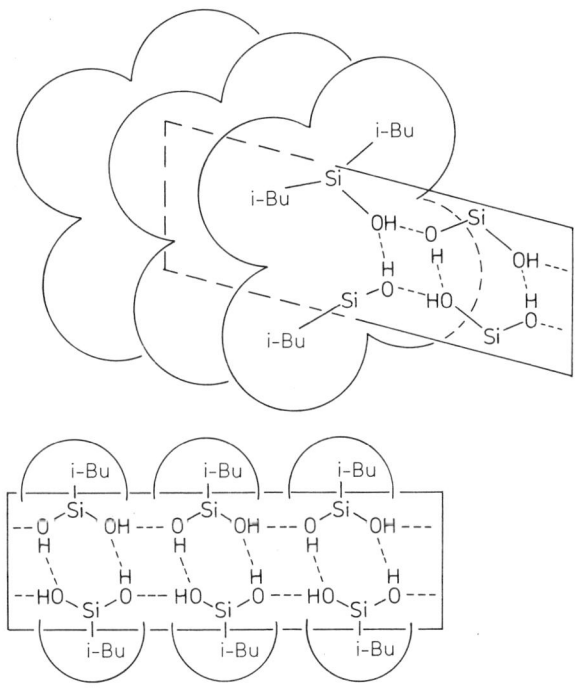

The links between the dimeric units are assumed to be broken at the transformation from the solid to the mesophase. It is worthwhile noting that the mesophase formed from the isotropic melt appears first as bright bands which resemble those observed in PDES. Earborn and Hartsborn [71] regarded these bands as "rodlike" structures and referred to these bands as "battonets". On the other hand, these structures can be interpreted as disclination lines [72]. From our point of view this question needs further consideration with allowance made for the fact, observed by Earborn and Hartshorn, that "battonets" can develop from striated bands arising after melting of acircular crystals and for the fact that the mesophase is not liquid, but a semi-solid phase.

The next example of thermotropic mesophase siloxane is octaphenylcyclotetrasiloxane (OPCTS). When OPCTS is heated up from room temperature, the following thermal transitions are observed [73]: a solid-solid transition at about 76°C, a solid-mesophase transition at about 188 °C, and transition into the isotropic melt at 205 °C. Based on the facts that the mesophase is optically isotropic and molecules are globular rather than elongated, it was suggested that the mesophase is a plastic crystalline phase. Subsequent calorimetric, optical and mechanical studies as well as results of X-ray diffraction, NMR and quasi-elastic neutron scattering (see pertinent references in [74]) supported the existence of the mesophase. However, analysis of powder neutron diffraction spectra of OPCTS has led Valino and Dianoux [75] to the conclusion that the mentioned mesophase resembles rather a smectic phase or discotic mesophase than a plastic crystal. It was found that molecules are arranged with long-range translational order in the direction perpendicular to the disc-like molecules of OPCTS (having "chair" isomer conformation) of the silicon ring and disordered (or short-range order) in the direction parallel to the discs. Note also a very interesting fact that the X-ray diffractogram of the mesophase of OPCTS is identical with that of PDPhS [26].

Lately, thermotropic mesophases have also been found in hexakis(trimethylsiloxy) disiloxane $\{[(CH_3)_3SiO]_3Si\}_2O$ [76] and in 1,3-dioxytetramethyldisiloxane (DOTES) [77] and 1,3-dioxytetra-n-propyldisiloxane (DOTPS) [77]. The former is interesting in that the methyl substituents are not bulky. The transition to a mesophase state occurs in the temperature range 233–245 K with the thermal effect 8.4 J/g. This mesophase is isotropic. Above 373 K this substance is able to flow but the crystalline reflexes in X-ray pattern remain below 500 K. It has been suggested that this mesophase can be interpreted as a plastic crystal.

The behaviour of DOTES and DOTPS is interesting in connection with the mesomorphic state of PDES and PDPS. It has been found that both molecules $HO[Si(C_2H_5)O]_2H$ and $HO[Si(C_3H_7)O]_2H$ can exist in a dimeric form resulting from the intermolecular hydrogen bonding and that these dimers melt step by step transforming into the isotropic melt via an intermediate mesomorphic state. DOTES is transformed into the mesophase state at 237 K and this transition is accompanied by a heat effect of 37.5 J/g and DOTPS is transformed into the mesophase at 317 K with a heat effect equal to 29.3 J/g. Transformation to the isotropic melt occurs at 310 and 346 K with heat effects 11.7 and 20.9 J/g, respectively, for DOTES and DOTPS. The X-ray pattern of DOTES consists of three reflexes corresponding to a hexagonal packing with the lattice parameter 11.1 Å. A similar X-ray profile was found also for DOTPS but the lattice parameter was a little larger. The optical textures of these substances in the mesophase state are similar to that of i-$Bu_2Si(OH)_2$. It has been suggested that the mesophase in DOTES is formed by dimeric cylindrical molecules with the diameter 11.1 Å and the length of about 9 Å, and that the mesophase resembles a discotic mesophase in that respect that two-dimensional long range order in packing of the cylinders in the direction perpendicular to their axis exists but translational order along their axis is absent.

The above examples do not seem to exhaust the list of low molecular organosiloxanes which are able to exist in a mesomorphic state. Further studies initiated at present by the discovery of the mesomorphic state in the flexible polyorganosiloxanes will lead to the discovery of such new substances in the near future.

3 Mesophases in Polyphosphazenes

3.1 General Description

The poly[organophosphazenes] or polyphosphazenes for short, are a rather new class of polymers of the general chemical structure

$$\left[\begin{array}{c} R \\ | \\ P-N \\ | \\ R \end{array} \right]_n$$

where R is OAlk, OAr, NHAlk, NHAr, Ar and other substituents. The synthesis and properties of polyphosphazenes are described in detail in the literature [78-81]. In our review, only those polyphosphazenes will be considered which can exist in the mesomorphic state, i.e. the polyphosphazenes with the alkoxy and aryloxy substituents. For the first time, the existence of an ordered phase after melting of the crystalline phase was discovered by Allen et al. [82] in poly(bistrifluorethoxyphosphazene), and by Singler et al. [79,83,84] in the chlorosubstituted polyaryloxyphosphazenes. Desper and Schneider [12] have introduced the term "mesomorphic phase" for such ordered phases. A detailed review of the first studies concerning the structure of the mesomorphic and crystalline phases was published in 1978 [14]. Later studies in this field were continued and some new information was published. In this part of the article, an attempt is made to summarize briefly both the new and previous results in order to have the possibility of comparing in more detail the structure and properties of the polyorganosiloxanes and polyorganophosphazenes in the mesomorphic state.

Although the polyphosphazene chain should be rigid as a result of the effect of the polyconjugation, it is flexible. For example, polyphosphonitrile chloride $[-P(Cl)_2=N-]_n$ is able to undergo large rubber-like deformations [85] and its glass transition temperature T_g is equal to 210 K [86]. There are a number of theoretical concepts explaining the absence of the delocalization of the electron bonding and the low magnitude of the rotational barriers around the P—N bond in the polyphosphazene chain [87,88]. It has been suggested that the magnitude of the rotational barrier is governed solely by the steric interactions between substituents or main chain atoms [88]. According to the results of the conformational analysis, the magnitude of the rotational barrier in the polyphosphazene macromolecules is not high, and reaches, in particular, 3.38 kJ/mole and 21.8 kJ/mole for the trifluoroethoxy- and phenoxy side groups, respectively [89]. The values of the gyration radii of the unperturbed polyphosphazene coils also indicate their high flexibility [79]. Hence, in spite of the absence of the rigorous quantitative characteristics of the flexibility of the polyphosphazene chains there is sufficient indirect evidence supporting their high flexibility.

3.2 Transitions

The polyphosphazenes are semi-crystalline polymers with the crystallinity depending on the conditions of the sample preparation. The glass transition temperature of the

Table 6. Transition temperatures, decomposition temperatures and heats of transitions for various polyphosphazenes $-[P(R_2)=N-]_n$ [14]

R	T_g (K)	T_m (K)	T (K)	T_d	ΔH_m, J/g
CF_3CH_2O	207	365	513	633	35.9
C_6H_5O	279	433	663	653	41.8
p-FC_6H_4O	259	442	618	—	45.1
m-ClC_6H_4O	249	363	643	653	24.2
p-ClC_6H_4O	277	442	638	683	27.6
m-$CH_3C_6H_4O$	248	363	621	623	34.7
p-$CH_3C_6H_4O$	273	425	613	583	15.5
m-$C_6H_5CH_2C_6H_4O$	270	382	—	320	43.5
$3,4$-$(CH_3)_2C_6H_3O$	268	369	598	588	19.2
$3,5$-$(CH_3)_2C_6H_3O$	279	340	593	613	5.0

amorphous phase depends on the type of substituents on P atoms. Table 6 summarizes the values of T_g for some mesomorphic polyphosphazenes. The twostep transition of the crystalline phase into the isotropic melt via an intermediate mesomorphic phase has been established by various methods. Schneider et al. [14,84] denoted the temperature of the crystalline phase-mesophase transition as T(1) and the temperature of the mesophase-isotropic melt transition as T_m. However, we will use below the definition introduced for the description of transitions in polyorganosiloxanes: the melting temperature, T_m, corresponds to the transition temperature of the crystalline phase into the mesophase and the isotropization temperature, T_i, refers to the transition of the mesophase into the isotropic melt. The values of the transition temperatures are listed in Table 6. It is interesting that the T_g/T_i ratio is about 0.4 as in the case of polyorganosiloxanes. The values of T_i are very close for various polyaryloxyphosphazenes, and in some cases even higher than the initial temperature of thermal degradation. This circumstance hampers the study of the transition to the isotropic state. In this respect, PBFP is the most suitable poly-

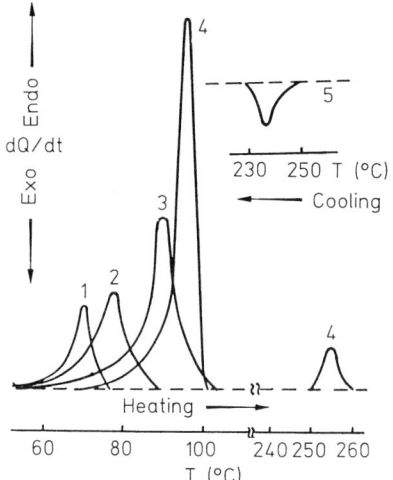

Fig. 27. DSC traces of PBFP samples with different thermal history [91]. 1 — original precipitated sample; 2 — sample annealed at 85 °C for 1 hour; 3 — sample annealed at 220 °C for 1 hour; 4 — sample annealed at 265 °C for 10 min; 5 — DSC trace of mesophase formation on cooling sample 4

phosphazene because its T_i is more than 100 K below the beginning of degradation. Quite naturally, this polyphosphazene has been studied most thoroughly. Two transitions were found in this polymer (Fig. 27). It has been established that T_m and heat of fusion ΔH_m depend on the conditions of the sample preparation and the thermal history [14,84,90,91]. The curves presented in Fig. 27 demonstrate it quite obviously. The reason for such behaviour will be discussed later, now we would only point out that the maximum values of T_m and ΔH_m are observed for the samples prepared by slow cooling of the isotropic melt.

T_m and ΔH_m listed in Table 6 for PBFP obtained for the samples with a wide MMD (in many cases $\bar{M}_w/\bar{M}_n > 10$) [92]. T_i for similar samples is 515 K and the heat of isotropization is 3.4 J/g. Lately, a new method of the synthesis of polyphosphazenes with a narrow MMD ($\bar{M}_w/\bar{M}_n < 1.6$) has been developed [93]. Samples of PBFP with such a MMD have $T_m = 371$ K, $\Delta H = 35.5$ J/g, $T_i = 528$ K, and $\Delta H_i = 3.8$ J/g [91]. Table 6 also shows the values of ΔH_m for some other polyphosphazenes. Unfortunately, the data presented are difficult to compare due to lack of information concerning the thermal history of some samples and the values of the heat of isotropization.

Dilatometric studies [84,94] have shown that both transitions in PBFP are accompanied with a considerable volume change (Fig. 28). Examination of these data shows that values of the relative volume changes during the melting and isotropization are both approximately 6%. Similar results were also obtained for poly(bis-m-chlorophenoxyphosphazene): the volume change accompanying the melting is 3.5% and the isotropization — 5.7% [84].

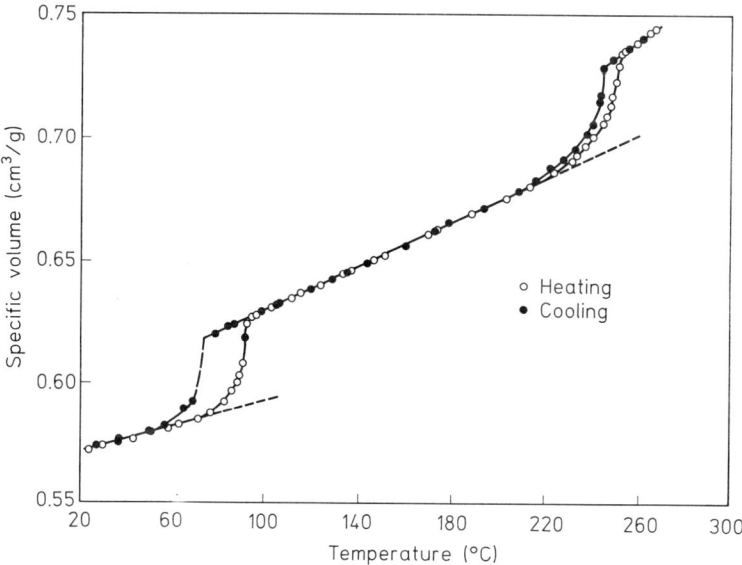

Fig. 28. Specific volume vs temperature plot for PBFP [94]. The volume coefficient of thermal expansion: 2.48×10^{-4} °C^{-1} in crystalline state (below 60 °C); 6.99×10^{-4} °C^{-1} in the mesomorphic state; 9.25×10^{-4} °C^{-1} in the molten state

The melting and isotropization of polyphosphazenes are accompanied by a considerable change of the mechanical properties. Above T_i, the PBFP is a isotropic viscous liquid and in the mesomorphic state the polymer resembles the weak paraffins. It is able to flow under pressure. Dynamic mechanical measurements [90,94,95] show that the modulus of elasticity, E', and "instantaneous" compliance, J_A, change in the transition region from the crystalline phase to the mesophase approximately 5-fold. Thus, films of PBFP become brittle and friable after heating through T_i. Similarly to the calorimetric behaviour, the mechanical properties depend strongly upon the thermal history and preparation conditions of the samples. The reason for this effect will be discussed in connection with the morphology of polyphosphazenes.

3.3 Structure of Crystalline Phases and Mesophases

The crystalline structure of polyphosphazenes attracted a considerable attention in the literature [14,96-102] because of the close relation between the crystalline and the mesomorphic structure. The data available for the crystal structure of some polyphosphazenes forming the mesomorphic phases are summarized in Table 7. As can be seen, all the crystals have a monoclinic or orthorombic structure. The fiber repeat period of most polyphosphazenes is about 4.9 Å which corresponds to the planar chain conformation close to the cis-trans type (Fig. 29). The exceptions are the polymers with bulky side groups, namely poly[bis(4-isopropylphenoxy)phosphazene] [103] and poly[bis(3,4-dimethylphenoxy)phosphazene] [104]. It was suggested that the interaction of bulky side groups of these polymers results in the displacement from the plane conformation of the polyphosphazene backbone. For the former macromolecule, a 3_1 helical conformation of the chain was proposed for the α-polymorph. The doubling of the fiber repeat for its β-polymorph and for the latter polymer was accounted for by a (cis-trans)$_4$ and (cis-trans-trans-trans)$_2$ conformation, respectively. It is noteworthy that in the cis-trans chain conformation all P—O bonds are directed in one direction and, therefore, each chain possesses directionality. The directionality of the phosphazene chains leads to the possibility of their various packing in the same crystal [96,102].

It has been found that some of the polyphosphazenes can exist in several crystalline polymorphs which only differ slightly in the packing of their chains. However, unlike PDES and PDPS, low temperature and high temperature polymorphs in polyphosphazenes have not been found yet. Which kind of crystal structure modifications occur during the crystallization depends on the molecular weight, the

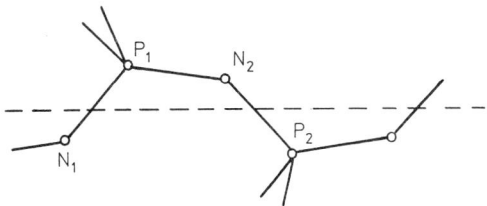

Fig. 29. Schematic drawing of the planar cis-trans conformation of polyphosphazene chain [14]

sample preparation conditions and its thermal history. In fact, the solution cast films of PBFP possess the α-orthorombic crystal structure, unlike the samples obtained by slow cooling from above T_m, which crystallize in the γ-orthorhombic form [98]. The crystallization of the low molecular weight PBFP from dilute solutions leads to the formation of the monoclinic β-modification [99, 100]. The crystal modification which occurs in the crystallization of poly[bis(p-fluorophenoxy)polyphosphazene] [101] and poly[bis(m-chlorophenoxy)phosphazene] [12] depends on the temperature attained on heating the mesomorphic polymers before crystallization, whereas the crystalline form of poly[(4-isopropylphenoxy)phosphazene] is determined by the rate of cooling from the mesomorphic state [105].

Heating of all mesophase polyphosphazenes above their melting points T_m is accompanied by the disappearance of the crystalline reflections from the X-ray diffraction pattern. However, above T_m the characteristic sharp reflection at small

Table 7. Crystalline structure of polyphosphazenes

Polymer	Unit cell (Å)	Refs.
$[N=P(OCH_2CF_3)-]_n$	α-form, orthorombic a = 10.14, b = 9.35, c = 4.86 Z = 2 β-form, monoclinic a = 10.03, b = 9.37, c = 4.86 γ = 91° γ-form, orthorombic a = 20.60, b = 9.40, c = 4.86 Z = 4	[98]
$[-N=P(OC_6H_5)_2]_n$	monoclinic a = 16.6, b = 13.8, c = 4.91 γ = 83, Z = 4	[99]
$[N=P(OC_6H_4\text{-}n\text{-}Cl)_2]_n$	orthorombic a = 13.08, b = 20.23, c = 4.90 Z = 4	[96]
$[N=P(OC_6H_4\text{-}m\text{-}Cl)]_n$	orthorombic c = 4.87	[14]
$[N=P(OC_6H_3\text{-}2,4\text{-}Cl_2)]_n$	monoclinic a = 21.6, b = 16.5, c = 4.86 γ = 94°, Z = 4	[100]
$[N=P(OC_6H_5\text{-}n\text{-}F)_2]_n$	α-form disordered β-form β-form, monoclinic a = 18.9, b = 13.2, c = 4.90 γ = 77	[101]
$[N=P(OC_6H_4\text{-}4\text{-}iC_3H_9)_2]_n$	α-form, monoclinic a = 24.55, b = 19.56, c = 5.56 γ = 101,0°, Z = 6 β-form, monoclinic a = 33.24, b = 22.69, c = 9.80 γ = 109.2°, Z = 16	[103]
$\{N=P[OC_6H_4\text{-}3,4\text{-}(CH_3)_2]_2\}_n$	orthorombic a = 15.85, b = 19.43, c = 9.85 Z = 2	[104]

angles and a diffuse halo with the center at 20 to 22° are observed for all samples; Fig. 30 demonstrates it quite obviously for PBFP [106]. In X-ray fiber patterns of a mesomorphic sample of PBFP, the first sharp reflection (11 Å) is concentrated at the equator indicating the existence of the lateral order in the packing of macromolecules above T_m [14, 82]. More detailed studies of the structure of poly[bis(p-chlorophenoxy)phosphazene] and poly[bis(m-chlorophenoxy)phosphazene] revealed two (or three) equatorial lines [12]. From the fact that the two sharp reflection are closely related to a hexagonal packing the structure of these polymers was referred to as pseudohexagonal. Diffuse meridianal scattering in the region 4.1 to 4.5 Å was attributed to (hkl) reflections broadened by rotational or longitudinal disorder. Hence, the structure of the mesomorphic state in the polyphosphazenes might be described as the two-dimentionally ordered packing of macromolecules in the plane normal to the chain axes. A similar structure was also suggested for PBFP, poly[bis(3,4-dimethylphenoxy)phosphazene] and poly[bis(benzylphenoxy)phosphazene] although they display only one sharp equatorial line in the mesomorphic state. This line was identified with the pseudohexagonal (100) reflection [14, 82]. Recently the pseudohexagonal structure was established also for the mesomorphic state of poly[bis(p-fluorophenoxy)phosphazene] [101].

The transformation from the orthorhombic or monoclinic crystals of polyphosphazenes listed in Table 7 to the hexagonal structure needs only a small changes in

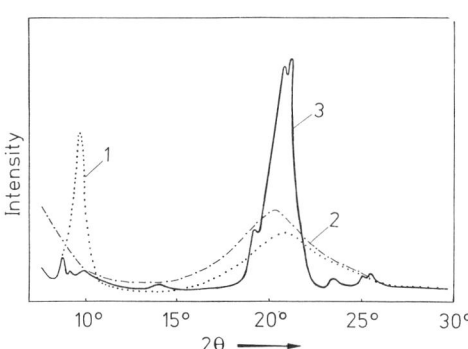

Fig. 30. Diffractograms of PBFP in different phase state [106]. 1 — mesomorphic state (384 K); 2 — isotropic molten state (533 K); 3 — crystalline state (298 K)

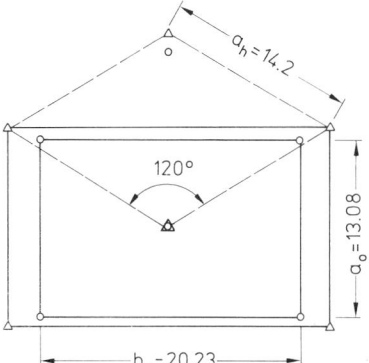

Fig. 31. Scheme of the transformation of an orthorombic unit cell into an pseudohexagonal unit cell [96]

the ab plane as can be seen from Fig. 31 illustrating the transformation of the orthorombic structure to the pseudohexagonal one in poly[bis(p-chlorophenoxy)-phosphazene] [96]. The mechanism of the structural transformation involves expansion of the orthorombic lattice by 9% in the direction a, and 20% in the direction b, leading to the alternation of the angle between the diagonals to 120°. A similar mechanism of the transformation of the orthorombic lattice to the hexagonal was also suggested for PBFP [98]. Although, according to the this reasoning, the pseudohexagonal structure of the mesophase in polyphosphazenes appears to be reasonable enough, we feel that for a final conclusion it is necessary to compare the theoretical and experimental densities of the mesophase as was done for PDES.

The most probable reason for the occurrence of the hexagonal type packing in polyphosphazenes consists in the appearance of dynamic conformational disorder in the form of a rapid backbone and side groups rotation that causes the molecular symmetry to be close to cylindrical [14]. Broad-line NMR study of the molecular motion in PBFP confirmed that the transition from the crystalline to the mesomorphic state is accompanied with the development of the rotational motion [107]. Pulsed NMR measurements show that the frequencies of the rotation are within 10^9 to 10^{10} s^{-1} [108]. The pseudohexagonal packing seems to exist in the whole temperature range where the mesophase exists. The interchain spacing changes reversibly with temperature as one can see from the change of (100) reflection (Fig. 32). The linear thermal expansion coefficient along the a-axis determined from similar data in the temperature range 380 to 485 K for PBFP samples crystallized upon slow cooling of the isotropic melt is 2.7×10^{-4} K^{-1} and the volume thermal expansion coefficient is 7×10^{-4} K^{-1} [94] (see also Fig. 28).

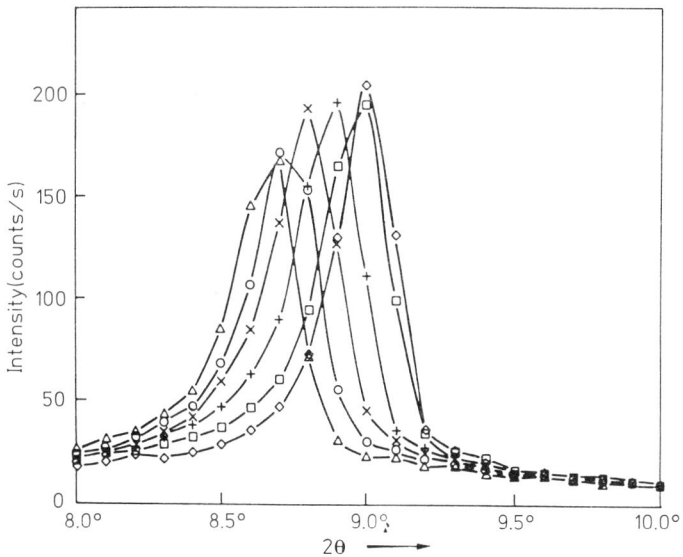

Fig. 32. Temperature dependence of 1-nm mesomorphic line for meltcrystallized PBFP [106]. □ — 100 °C; + — 135 °C; × — 170 °C; ○ — 200 °C; △ — 230 °C;   — 100 °C

3.4 Morphology of Crystalline and Mesomorphic State

The morphologies of the crystalline state and the mesomorphic state are closely related. Both depend on the sample preparation conditions and thermal history. Defects in the structure of macromolecules also influence the morphology [109]. Crystalline films of polyphosphazenes obtained by solution casting, PBFP in particular, exhibit the spherulitic morphology which remains in the mesophase state after melting of the crystalline structure, and it disappears only after complete isotropization at T_i [84, 94, 98, 109, 110]. On the other hand, upon cooling of the isotropic melt the rod-like optically anisotropic mesomorphic regions develop, which upon subsequent crystallization are transformed into the crystals of the same shape [84, 98]. Similar optical texture is observed in films of PBFP casted from solution on hot liquid surfaces above T_m, whereas films of poly(bisphenoxyphosphazene) prepared under similar conditions display a highly developed lamellar structure [111]. Crystalline spherulites have a radial structure characterized by negative birefringence: the molecular chains are oriented predominantly perpendicular to the radius of the spherulite with the crystallographic "a" direction of the unit cell lying along the radius [94, 110]. In the mesomorphic state, the birefringence of the spherulites remains negative. It means that the orientation of macromolecules in space is unchanged, too. However, the magnitude of the negative birefringence increases with temperature. It is noteworthy that this process is accompanied with a simultaneous irreversible increase in the intensity of the (100)-reflection in diffractograms [106]. All these facts show that on heating, the structure of the mesomorphic state becomes more ordered. This ordering, in turn, influences the properties of the polymer in the crystalline state after the transformation of the mesomorphic sample into the crystalline state on cooling. Macroscopically this leads to the already mentioned strong dependence of the mechanical properties of the films on their thermal history and clearly manifests itself as an increase in the melting point and the heat of fusion of the crystalline phase [84, 90, 91]. Figure 33 shows the dependence of T_m and ΔH_m on the temperature and time of the exposition in the mesomorphic state. The results demonstrate that the ordering processes proceed rather fast.

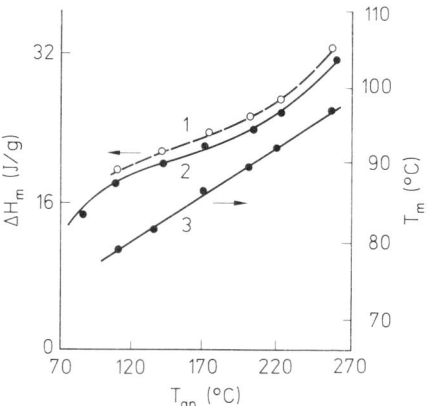

Fig. 33. Heat of fusion ΔH_m and melting peak temperature T_m of PBFP as functions of annealing temperature and annealing time [91]. Annealing time: 1 — 1 h; 2 — 3 s; 3 — 15 min

The reason for the influence of the mesomorphic state on the final properties of the crystal films becomes quite obvious after the comparison of the phase content in the mesomorphic and crystalline states. Such data were obtained with pulse NMR and DSC [91]. It was established that PBFP in the mesomorphic state consists of two phases. In reprecipitated samples and films above T_m, the mesomorphic and amorphous phases coexist in the metastable equilibrium which displaces to the mesophase formation with increasing temperature. Analysis of the relation of the mesomorphic, crystalline, and amorphous phases in the samples with a different thermal history revealed the following picture: upon cooling of the polymer from the mesomorphic state, apparently, only the mesomorphic regions crystallize and correspondingly the final degree of crystallinity is approximately equal to the content of the mesophase which irreversibly increases with the annealing temperature. The mechanism of the morphological changes in the polymer in the mesomorphic state may be described as follows: The crystallization of polyphosphazene from a solution during evaporation of solvents proceeds without a preliminary formation of the mesophase because of a considerable decrease of the upper temperature limit of its existence in the presence of the solvent. Upon such nonequilibrium crystallization, the relatively small crystallites connected by tie-molecules form. It was suggested that the α-orthorhombic crystals with the folded-chain macroconformation occur during such crystallization [98]. Initial crystallinity of samples obtained under these conditions is not high. Above T_m after melting of crystallites, the tie-molecules which compose the amorphous phase of the system connect the mesomorphic regions. These regions are metastable but their reorganization on further heating may occur when the system reaches an appropriate temperature at which the mesomorphic regions "melt" and which is dependent on their dimensions. As a result of such "melting", large parts of macromolecules gain enough mobility to become reorganized into the mesomorphic domains of larger dimensions. One of the possible mechanisms of the organization seems to be the increase in fold length, which has been established experimentally in polyphosphazene single crystals above T_m [100]. Thus, the heating cycle is accompanied by the subsequent increase in the dimensions of the reorganizing mesomorphic regions and the content of the mesophase. Just before the isotropization the content of the mesophase reaches very high values (of the order 0.9) and, therefore, all samples irrespective of their thermal history and preparation conditions have very close values of T_i and ΔH_i. The reorganized mesophase region crystallize into the more extended chain crystals (γ-orthorombic crystals [98]), and as a result they have higher values of T_m. Recently, an increase in crystallinity and an improvement of crystal perfection in poly[bis(p-methylphenoxy)phosphazene] after heating above T_m has also been established with SAXS and WAXS techniques using synchrotron radiation [112]. Hence, like in PDES and PDPS a very pronounced effect of the preliminary formation of the mesophase on the crystallization of polyphosphazenes exists.

Unlike PDES, PBFP could not be quenched into the completely amorphous state. The reason of this fact seems to consist in a very large difference between T_m and T_i. Upon slow cooling, the isotropic melt is transformed into the mesophase at 10 to 15 K below T_i (see Fig. 27). The kinetics of the formation of the mesophase has not yet been sufficiently studied. Nevertheless, by analogy with PDES one can suggest that this process involves the nucleation and subsequent growth of nuclei and the criteria of the stability of growing mesophase domains force the macromolecules to be completely incorporated. This suggestion follows from the practically complete transition of the isotropic melt into the mesophase and the subsequent almost complete crystallization of the polymer as well (the degree of crystallinity reaches 95%). Such very high crystallinity is a reason for the brittleness of polyphosphazene samples after heating above T_i.

The crystallization kinetics of the mesomorphic polyphosphazenes has been studied insufficiently and more study is required. In particular, the dilatometric results of the crystallization kinetics of the mesophase sample of PBFP [94] and X-ray data of crystallization of poly(p-methylphenoxypolyphosphazene) [112] are of a preliminary character and cannot be used for final conclusions concerning the specific features of the process. It is necessary to have many more experimental results before the effect of the mesophase on crystallization may be evaluated systematically.

4 Concluding Remarks

A number of organo-element polymers form stable ordered phases. These phases can be treated as mesomorphic phases in that they are an intermediate level of organization between the crystalline and isotropic amorphous phases. They possess two-dimensional order in the plane perpendicular to their longitudinal axis. Macromolecules in these phases display a rather high level of molecular mobility, i.e. rotational motion and translational displacement. The structure and molecular dynamics of these phases are close to that of high temperature crystalline modifications of some organic polymers [1, 14, 54] and low molecular substances [113]. However, formation of these thermotropic mesophases in element-organic polymers cannot be considered as a unique ability of these macromolecules only. It seems to be a general feature of element-organic compounds, at least organosiloxanes. All these mesophases differ from classical liquid crystals and for their description the term "disordered crystal" (including condis crystal as a particular case) seems to be more appropriate.

First of all, these phases are not liquid but semisolid, and in this respect they resemble plastic crystals, although, from the structural point of view they differ from plastic crystals by the absence of not the rotational order, but by the translational order in one or two directions as well as azimuthal order (or conformational disordering) in polymers.

The nature of the thermotropic mesophases in flexible element-organic polymers seems also to be different from the nature of liquid crystals, first of all, by a more important role of the energy factor in forming these mesophases. In principle, the transformation of a crystal into the isotropic melt through intermediate states means the existence of a hierarchy of discrete intermolecular interactions which are gradually released with increasing temperature and this gives rise to the appearance of a disordered crystal. One can probably suggest the existence of some specific interactions between element-organic molecules resulting from the polarity of the siloxane and phosphazene bonds and from possible additional dipole-dipole interactions, in particular, owing to a different electronegativity of Si and C atoms. However, the fact that not all substances with identical side substituents on Si atoms form thermotropic mesophases (for example, OPCTS and DECPS form mesophases, but other cycles of these homologous series do not form them; similarly, DODTS forms a mesophase and DESD does not) shows that for the realization of the specific intermolecular interaction leading to the formation of a mesophase, a corresponding geometrical configuration of molecules is necessary.

The influence of the molecular configuration on mesophase formation is clearly seen for cyclolinear polyorganosiloxanes, for which the enhanced rigidity leads to an increased configurational anisotropy of macromolecules which, causes widening of the temperature interval for the existence of a mesophase. The rigidity of polyorganosiloxanes can be confered by both the incorporation of bulky substituents and the change of the structure of the cyclolinear backbone.

The most rigid chains are those of ladder polyorganosilsesquioxanes. Together with esters of cellulose, these polymers are an example of the lyotropic liquid crystalline system which is formed by semirigid macromolecules. The structure of their solid mesophasese has not been studied sufficiently enough, therefore, at present it is not clear whether it can be considered as a disordered crystal or as a liquid crystalline phase. Further study of the structure of the mesophases in element-organic polymers and low molecular substances would undoubtedly be very useful for the development of the concept of disordered crystals.

5 References

1. Smith GW (1975) Adv. Liquid Crystals Vol. 1, 3. Academic, New York
2. Wunderlich B, Grebowicz J (1984) Adv Polym Sci Vol. 60/61, 1. Springer, Berlin Heidelberg New York
3. Papkov SP, Kulichihin VG (1977) Liquid Crystalline State of Polymers (in Russian) Khimija, Moscow
4. Flory PI (1984) Adv Polym Sci, Vol 59, p 1. Springer, Berlin Heidelberg New York
5. Shibaev VP, Plate NA (1984) Adv Polym Sci, Vol 60/61, 173. Springer, Berlin Heidelberg New York
6. Finkelmann H, Rehage G (1984) Adv Polym Sci, Vol 60/61, p 99. Springer, Berlin Heidelberg New York
7. Ober CH, Jin J, Zhou O, Lenz RW (1984) Adv Polym Sci, Vol 59, p 103. Springer, Berlin Heidelberg New York
8. Yeh GSY (1972) Pure Appl Chem 31: 65
9. Zachmann G (1975) Pure Appl Chem 43: 207
10. Wendorff JH (1978) in: Blumstein A (ed) Liquid Crystalline Order in Polymers. Academic, New York
11. Tswetkov VN, Rjumzev EI, Shtennikova IN (1978) in: Liquid Crystalline Order in Polymers. Blumstein A (ed). Academic, New York
12. Desper CR, Schneider NS (1976) Macromolecules 9: 424
13. Beatty CL, Pochan JM, Froix MF, Hinman DF (1975) Macromolecules 8: 547; Pochan JM, Beatty CL, Hinman DF (1975) J Polym Sci Polym Phys Ed 13: 977
14. Schneider NS, Desper CR, Birs JJ (1978) in: Liquid Crystalline Order in Polymers, Blumstein A (ed) Academic, New York
15. Kim C, Allcock HR (1987) Macromolecules 20: 1726
16. Singler RE, Willingham RA, Lenz RW, Furukawa A, Finkelmann H (1987) Macromolecules 20: 1727
17. Neilson R, Hani R, Neilson P-W, Meister JJ, Roy AK, Hagnauer CL (1987) Macromolecules 20: 910
18. Flory PJ (1969) Statistical Mechanics of Chain Molecules. Wiley, New York

19. Lee CL, Emerson EA (1967) J Polym Sci A-2,5: 829
20. Mark JE, Chin DS, Su TK (1978) Polymer 19: 407
21. Lee CL, Johanson OK, Flaningan OL, Hahn P (1969) Am Chem Soc, Polym Preprints 10: 1319
22. Godovsky YK, Papkov VS (1986) Makromol Chem, Macromol Symp 4: 71
23. Godovsky YK, Makarova NN, Papkov VS, Kuzmin NN (1985) Vysokomol Soedin B27: 164
24. Papkov VS, Godovsky YK (1985) 2nd FRG-USSR Symposium on Macromolecules, Abstracts, p 19
25. Godovsky YK, Mamaeva II, Makarova NN, Papkov VS, Kuzmin NN (1985) Makromol Chem, Rapid Commun 6: 797
26. Tsvankin DY, Levin VY, Papkov VS, Zhukov VP, Zhdanov AA, Andrianov KA (1979) Vysokomol Soedin A21: 2126
27. Papkov VS, Godovsky YK, Svistunov VS, Litvinov VM, Zhdanov AA (1984) J Polym Sci, Polym Chem Ed 22: 3617
28. Papkov VS, Godovsky YK, Litvinov VM, Svistunov VS, Zhdanov AA (1981) 4-th International Conference on Liquid Crystals, Tbilisi, Proc., Vol 2, p 210
29. Papkov VS, Godovsky YK, Svistunov VS, Litvinov VM, Zhdanov AA (1982) First Allunion Symposium on Polymer Liquid Crystals, Suzdal, Preprints p 108
30. Tsvankin DY, Papkov VS, Zhukov VP, Godovsky YK, Svistunov VS, Zhdanov AA (1985) J Polym Sci, Polym Chem Ed 23: 1043
31. Litvinov VM, Lavruchin BD, Papkov VS, Zhdanov AA (1985) Vysokomol Soedin A27: 1529
32. Froix MF, Beatty CL, Pochan JM, Hinman DD (1975) J Polym Sci, Polym Phys Ed 13: 1269
33. Litvinov VM, Lavruchin BD, Papkov VS, Zhdanov AA (1983) Dokl Akad Nauk SSSR 271: 900
34. Litvinov VM, Lavruchin BD, Papkov VS, Zhdanov AA (1982) First Allunion Symposium on Polymer Liquid Crystals, Suzdal, Preprints p 109
35. Papkov VS, Svistunov VS, Godovsky YK, Zhdanov AA (1987) J Polym Sci, Polym Phys. Ed 25: 1858
36. Wunderlich B (1976) Macromolecular Physics, Vol 2. Academic, New York
37. Pochan JM, Hinman DF, Froix MF (1976) Macromolecules 9: 611
38. Shulgin AI, Godovsky YK. Vysokomol Soedin (in press)
39. Papkov VS, Svistunov VS, Godovsky YK. Vysokomol Soedin (in press)
40. Godovsky YK, Volegova IA, Valetskay LA. Vysokomol Soedin (in press)
41. Godovsky YK (1986) Adv Polym Sci 76: 31
42. Mark JE (1973) Rubber Chem Techn 46: 593
43. Shibanov YD, Radzhabov TM, Komaricheva LI, Godovsky YK (1987) Composite Polymer Materials, N32: 28
44. Makarova NN, Petrova IM, Godovsky YK, Lavruchin BD, Zhdanov AA (1983) Dokl Akad Nauk SSSR 269: 1368
45. Makarova NN, Petrova IM, Godovsky YK, Zhdanov AA (1984) Bull Izobr N44
46. Makarova NN, Godovsky YK, Kuzmin NN (1987) Makromol Chem 188: 119
47. Mamaeva II, Pavlova SA, Tverdochlebova II, Makarova NN. Vysokomol Soedin (in press)
48. Godovsky YK, Makarova NN, Petrova IM, Zhdanov AA (1984) Makromol Chem, Rapid Commun 5: 427
49. Godovsky YK, Makarova NN, Petrova IM, Zhdanov AA (1984) Abstracts from the 5. Int. Liquid Crystal Conference of Socialist Countries, Odessa, USSR, Vol 2, Part II, p 118
50. Makarova NN, Godovsky YK, Zhiganshina RI (1985) Abstracts of XXII. USSR Conference on High Polymers, Alma-Ata, Proc., p 89
51. Makarova NN, Godovsky YK, Petrova IM (1986) Abstracts of YI USSR Conference on Organosilicon Compounds, Riga, p 46
52. Godovsky YK, Makarova NN, Mamaeva II (1986) Makromol Chem 7: 325
53. Godovsky YK, Makarova NN, Kuzmin NN. Vysokomol Soedin (in press)
54. Godovsky YK, Papkov VS (1988) in: Mesomorphic State of Flexible Polymers, Khimija, Moscow
55. Godovsky YK, Makarova NN, Kuzmin NN (1987) 2nd USSR Symposium on Polymer Liquid Crystals, Suzdal

56. Makarova NN, Godovsky YK (1986) Vysokomol Soedin B28: 243
57. Brown JF (1963) J Polym Sci C-1: 83
58. Andrianov KA, Makarova NN (1970) Vysokomol Soedin A12: 663
59. Tsvetkov VN, Andrianov KA, Makarova NN, Vitovskay MG, Rjumzev EI (1973) Europ Polym J 9: 27
60. Papkov VS (1979) Doctor dissertation, INEOS AN SSSR, Moscow
61. Andrianov KA, Slonimsky GL, Genin YV, Gerasimov VI, Levin VY (1969) Dokl Akad Nauk SSSR 187: 1285
62. Andrianov KA, Slonimsky GL, Tsvankin DA, Papkov VS, Levin VY (1974) Vysokomol Soedin B16: 208
63. Andrianov KA, Slonimsky GL, Zhdanov AA, Tsvankin DY, Levin VY, Papkov VS (1976) J Polym Sci A-1,14: 1205
64. Kwachew YP, Perepechko II, Papkov VS, Levin VY, Zhdanov AA (1973) Mechanika Polymerov N 5: 804
65. Papkov VS, Ilina MN, Kwachew YP, Makarova NN, Zhdanov AA, Slonimsky GL, Andrianov KA (1975) Vysokomol Soedin A17: 2050
66. Volynskaya AV, Godovsky YK, Papkov VS (1980) Vysokomol Soedin A21: 1039
67. Papkov VS, Obolonkova ES, Ilina MN, Zhdanov AA, Slonimsky GL, Andrianov KA (1980) A22: 117
68. Papkov VS, Ilina MN, Perzova NV, Makarova NN, Slonisky GL, Zhdanov AA, Andrianov KA (1977) Vysokomol Soedin A19: 2551
69. Hurd DT, Osthoff RC (1953) J Am Chem Soc 75: 234
70. Eaborn C (1952) J Chem Soc 2840
71. Eaborn C, Hartshorn NH (1955) J Chem Soc 549
72. Booning JD, Lydon JF, Eaborn C, Jackson PM, Goodby JM, Gray GW (1982) J Chem Soc Farad Trans I 78: 713
73. Keyes PH, Daniels WB (1975) J Chem Phys 62: 2000
74. Smith GW (1986) Mol Cryst Liquid Crystals 132: 385
75. Valino F, Dianoux A (1978) Ann Phys 3: 151
76. Matuchina EV, Kuzmin NN, Antipov EM, Molchanov BV et al. Vysokomol Soedin (in press)
77. Makarova NN, Kuzmin NN, Godovsky YK, Matuchina EV. Vysokomol Soedin (in press)
78. Allcock HR (1972) Chem Rev 72: 315
79. Singler RE, Schneider NS, Hagnauer GL (1975) Polym Eng Sci 15: 312; Hagnauer GL, Schneider NS (1972) J Polym Sci, Part A-2, 10: 669
80. Tur DR, Vinogradova SV (1982) Vysokomol Soedin 24: 2447
81. Vinogradova SV, Tur DR, Minosyanz II (1984) Uspechi Chimii 53: 87
82. Allen G, Lewis CJ, Todd SM (1970) Polymer 11:44
83. Singler RE, Hagnauer GL, Schneider NS, LaLiberto BR, Sacher RE, Matton RW (1974) J Polym Sci, Polym Phys Ed 12: 433
84. Schneider NS, Desper CR, Singler RE (1976) J Appl Polym Sci 20: 3087
85. Polymer Encyclopedia, Moscow: Soviet Encyclopedia, Vol 3, p 79 (1977)
86. Allcock HR, Kugel RL, Valan KJ (1966) Inorg Chem 5: 1709
87. Craig DP, Paddock NC (1971) in: Snyder JF (ed) Nonbenzenoid Aromatics, Academic, New York
88. Allcock HR, Allen RW, Meister JJ (1976) Macromolecules 9: 950
89. Allen RW, Allcock HR (1976) Macromolecules 9: 956
90. Choy JC, Maggil JH (1981) J Polym Sci, Polym Chem Ed 19: 2495
91. Papkov VS, Litvinov VM, Dubovik II, Slonimsky GL, Tur DR, Vinogradova SV, Korshak VV (1985) Dokl Akad Nauk SSSR 284: 1423
92. Hagnauer GL, LaLiberto BR (1976) J Polym Sci, Polym Phys Ed 14: 367
93. Tur DR, Timofeeva GI, Ilina MN, Gogyadze CY, Alikhanova N (1985) Abstracts of XXII. USSR Conference on Highpolymers, Alma-Ata, p 92
94. Masuko T, Simeone RL, Maggil JH, Plasek DJ (1984) Macromolecules 17: 2857
95. Conneley TM, Gillham JK (1976) J Appl Polym Sci 20: 479
96. Bishop SM, Hall JH (1974) Polymer J 6: 193
97. Allcock HR, Acrus RA, Stroh EG (1980) Macromolecules 13: 919

98. Kojima M, Maggil JH (1985) Makromol Chem 186: 649
99. Kojima M, Maggil JH (1983) Polym Commun 24: 329
100. Kojima M, Kenge W, Maggil JH (1984) Macromolecules 17: 1421
101. Matsuzawa S, Jamaura K, Wanigami T, Higuch M (1985) Colloid Polym Sci 263: 888
102. Giglo E, Pompa F, Ripamonti A (1962) J Polym Sci 59: 293
103. Meille SV, Porzio W et al. (1986) Makromol Chem, Rapid Commun 7: 217
104. Burkhort CW, Gillette PC, Lando JB, Beres JJ (1983) J Polym Sci, Polym Phys Ed 21: 2349
105. Meille SV, Porzio W, Balgnezi A, Glerio M (1987) Makromol Chem, Rapid Commun 8: 43
106. Russell TR, Anderson DP, Stein RS, Desper CR, Beres, Schneider NS (1984) Macromolecules 17: 1795
107. Alexander MN, Desper CR, Sagalyn PL, Schneider NS (1977) Macromolecules 10: 721
108. Litvinov VM, Papkov VS, Tur DR (1986) Vysokomol Soedin A28: 289
109. Ferrar WT, Marshall AS (1985) Am Chem Soc, Polym Preprints 26: 222
110. Kojima M, Maggil JH (1985) Polymer 26: 1971
111. Maggil JH, Petermann J, Rieck U (1986) Colloid Polym Sci 264: 570
112. Maggil JH, Rieckel C (1986) Makromol Chem, Rapid Commun 7: 287
113. Drotloff H, Emeis D, Waldron RF, Möller M (1987) Polymer 28: 1200

Editor: K. Dušek
Received November 24, 1987

Author Index Volumes 1–88

Allegra, G. and *Bassi, I. W.:* Isomorphism in Synthetic Macromolecular Systems. Vol. 6, pp. 549–574.
Andrade, J. D., Hlady, V.: Protein Adsorption and Materials Biocompability: A. Tutorial Review and Suggested Hypothesis. Vol. 79, pp. 1–63.
Andrews, E. H.: Molecular Fracture in Polymers. Vol. 27, pp. 1–66.
Anufrieva, E. V. and *Gotlib, Yu. Ya.:* Investigation of Polymers in Solution by Polarized Luminescence. Vol. 40, pp. 1–68.
Apicella, A. and *Nicolais, L.:* Effect of Water on the Properties of Epoxy Matrix and Composite. Vol. 72, pp. 69–78.
Apicella, A., Nicolais, L. and *de Cataldis, C.:* Characterization of the Morphological Fine Structure of Commercial Thermosetting Resins Through Hygrothermal Experiments. Vol. 66, pp. 189–208.
Argon, A. S., Cohen, R. E., Gebizlioglu, O. S. and *Schwier, C.:* Crazing in Block Copolymers and Blends. Vol. 52/53, pp. 275–334.
Aronhime, M. T., Gillham, J. K.: Time-Temperature Transformation (TTT) Cure Diagram of Thermosetting Polymeric Systems. Vol. 78, pp. 81–112.
Arridge, R. C. and *Barham, P. J.:* Polymer Elasticity. Discrete and Continuum Models. Vol. 46, pp. 67–117.
Aseeva, R. M., Zaikov, G. E.: Flammability of Polymeric Materials. Vol. 70, pp. 171–230.
Ayrey, G.: The Use of Isotopes in Polymer Analysis. Vol. 6, pp. 128–148.

Bässler, H.: Photopolymerization of Diacetylenes. Vol. 63, pp. 1–48.
Baldwin, R. L.: Sedimentation of High Polymers. Vol. 1, pp. 451–511.
Bascom, W. D.: The Wettability of Polymer Surfaces and the Spreading of Polymer Liquids. Vol. 85, pp. 89–124.
Balta-Calleja, F. J.: Microhardness Relating to Crystalline Polymers. Vol. 66, pp. 117–148.
Barbé, P. C., Cecchin, G. and *Noristi, L.:* The Catalytic System Ti-Complex/$MgCl_2$. Vol. 81, pp. 1–83.
Barton, J. M.: The Application of Differential Scanning Calorimetry (DSC) to the Study of Epoxy Resins Curing Reactions. Vol. 72, pp. 111–154.
Ballauff, M. and *Wolf, B. A.:* Thermodynamically Induced Shear Degradation. Vol. 84, pp. 1–31.
Basedow, A. M. and *Ebert, K.:* Ultrasonic Degradation of Polymers in Solution. Vol. 22, pp. 83–148.
Batz, H.-G.: Polymeric Drugs. Vol. 23, pp. 25–53.
Baur, H. see *Wunderlich, B*: Vol. 87, pp. 1–121.
Bell, J. P. see *Schmidt, R. G.:* Vol. 75, pp. 33–72.
Bekturov, E. A. and *Bimendina, L. A.:* Interpolymer Complexes. Vol. 41, pp. 99–147.
Bergsma, F. and *Kruissink, Ch. A.:* Ion-Exchange Membranes. Vol. 2, pp. 307–362.
Berlin, Al. Al., Volfson, S. A., and *Enikolopian, N. S.:* Kinetics of Polymerization Processes. Vol. 38, pp. 89–140.
Berry, G. C. and *Fox, T. G.:* The Viscosity of Polymers and Their Concentrated Solutions. Vol. 5, pp. 261–357.
Bevington, J. C.: Isotopic Methods in Polymer Chemistry. Vol. 2, pp. 1–17.
Beylen, M. van, Bywater, S., Smets, G., Szwarc, M., and *Worsfold, D. J.:* Developments in Anionic Polymerization — A Critical Review. Vol. 86, pp. 87–143.

Bhuiyan, A. L.: Some Problems Encountered with Degradation Mechanisms of Addition Polymers. Vol. 47, pp. 1–65.

Bird, R. B., Warner, Jr., H. R., and *Evans, D. C.:* Kinetic Theory and Rheology of Dumbbell Suspensions with Brownian Motion. Vol. 8, pp. 1–90.

Biswas, M. and *Maity, C.:* Molecular Sieves as Polymerization Catalysts. Vol. 31, pp. 47–88.

Biswas, M., Packirisamy, S.: Synthetic Ion-Exchange Resins. Vol. 70, pp. 71–118.

Block, H.: The Nature and Application of Electrical Phenomena in Polymers. Vol. 33, pp. 93–167.

Bodor, G.: X-ray Line Shape Analysis. A. Means for the Characterization of Crystalline Polymers. Vol. 67, pp. 165–194.

Böhm, L. L., Chmelíř, M., Löhr, G., Schmitt, B. J. and *Schulz, G. V.:* Zustände und Reaktionen des Carbanions bei der anionischen Polymerisation des Styrols. Vol. 9, pp. 1–45.

Bölke, P. see Hallpap, P.: Vol. 86, pp. 175–236.

Boué, F.: Transient Relaxation Mechanisms in Elongated Melts and Rubbers Investigated by Small Angle Neutron Scattering. Vol. 82, pp. 47–103.

Bovey, F. A. and *Tiers, G. V. D.:* The High Resolution Nuclear Magnetic Resonance Spectroscopy of Polymers. Vol. 3, pp. 139–195.

Braun, J.-M. and *Guillet, J. E.:* Study of Polymers by Inverse Gas Chromatography. Vol. 21, pp. 107–145.

Breitenbach, J. W., Olaj, O. F. und *Sommer, F.:* Polymerisationsanregung durch Elektrolyse. Vol. 9, pp. 47–227.

Bresler, S. E. and *Kazbekov, E. N.:* Macroradical Reactivity Studied by Electron Spin Resonance. Vol. 3, pp. 688–711.

Brosse, J.-C., Derouet, D., Epaillard, F., Soutif, J.-C., Legeay, G. and *Dušek, K.:* Hydroxyl-Terminated Polymers Obtained by Free Radical Polymerization. Synthesis, Characterization, and Applications. Vol. 81, pp. 167–224.

Bucknall, C. B.: Fracture and Failure of Multiphase Polymers and Polymer Composites. Vol. 27, pp. 121–148.

Burchard, W.: Static and Dynamic Light Scattering from Branched Polymers and Biopolymers. Vol. 48, pp. 1–124.

Bywater, S.: Polymerization Initiated by Lithium and Its Compounds. Vol. 4, pp. 66–110.

Bywater, S.: Preparation and Properties of Star-branched Polymers. Vol. 30, pp. 89–116.

Bywater, S. see Beylen, M. van: Vol. 86, pp. 87–143.

Candau, S., Bastide, J. und *Delsanti, M.:* Structural, Elastic and Dynamic Properties of Swollen Polymer Networks. Vol. 44, pp. 27–72.

Carrick, W. L.: The Mechanism of Olefin Polymerization by Ziegler-Natta Catalysts. Vol. 12, pp. 65–86.

Casale, A. and *Porter, R. S.:* Mechanical Synthesis of Block and Graft Copolymers. Vol. 17, pp. 1–71.

Cecchin, G. see Barbé, P. C.: Vol. 81, pp. 1–83.

Cerf, R.: La dynamique des solutions de macromolecules dans un champ de vitresses. Vol. 1, pp. 382–450.

Cesca, S., Priola, A. and *Bruzzone, M.:* Synthesis and Modification of Polymers Containing a System of Conjugated Double Bonds. Vol. 32, pp. 1–67.

Chiellini, E., Solaro, R., Galli, G. and *Ledwith, A.:* Optically Active Synthetic Polymers Containing Pendant Carbazolyl Groups. Vol. 62, pp. 143–170.

Cicchetti, O.: Mechanisms of Oxidative Photodegradation and of UV Stabilization of Polyolefins. Vol. 7, pp. 70–112.

Clark, A. H. and *Ross-Murphy, S. B.:* Structural and Mechanical Properties of Biopolymer Gels. Vol. 83, pp. 57–193.

Clark, D. T.: ESCA Applied to Polymers. Vol. 24, pp. 125–188.

Colemann, Jr., L. E. and *Meinhardt, N. A.:* Polymerization Reactions of Vinyl Ketones. Vol. 1, pp. 159–179.

Comper, W. D. and *Preston, B. N.:* Rapid Polymer Transport in Concentrated Solutions. Vol. 55, pp. 105–152.

Corner, T.: Free Radical Polymerization — The Synthesis of Graft Copolymers. Vol. 62, pp. 95–142.
Crescenzi, V.: Some Recent Studies of Polyelectrolyte Solutions. Vol. 5, pp. 358–386.
Crivello, J. V.: Cationic Polymerization — Iodonium and Sulfonium Salt Photoinitiators, Vol. 62, pp. 1–48.

Dave, R. see *Kardos, J. L.:* Vol. 80, pp. 101–123.
Davydov, B. E. and *Krentsel, B. A.:* Progress in the Chemistry of Polyconjugated Systems. Vol. 25, pp. 1–46.
Derouet, F. see *Brosse, J.-C.:* Vol. 81, pp. 167–224.
Dettenmaier, M.: Intrinsic Crazes in Polycarbonate Phenomenology and Molecular Interpretation of a New Phenomenon. Vol. 52/53, pp. 57–104.
Diaz, A. F., Rubinson, J. F., and *Mark, H. B., Jr.:* Electrochemistry and Electrode Applications of Electroactive/Conductive Polymers. Vol. 84, pp. 113–140.
Dobb, M. G. and *McIntyre, J. E.:* Properties and Applications of Liquid-Crystalline Main-Chain Polymers. Vol. 60/61, pp. 61–98.
Döll, W.: Optical Interference Measurements and Fracture Mechanics Analysis of Crack Tip Craze Zones. Vol. 52/53, pp. 105–168.
Doi, Y. see *Keii, T.:* Vol. 73/74, pp. 201–248.
Dole, M.: Calorimetric Studies of States and Transitions in Solid High Polymers. Vol. 2, pp. 221–274.
Donnet, J. B., Vidal, A.: Carbon Black-Surface Properties and Interactions with Elastomers. Vol. 76, pp. 103–128.
Dorn, K., Hupfer, B., and *Ringsdorf, H.:* Polymeric Monolayers and Liposomes as Models for Biomembranes How to Bridge the Gap Between Polymer Science and Membrane Biology? Vol. 64, pp. 1–54.
Dreyfuss, P. and *Dreyfuss, M. P.:* Polytetrahydrofuran. Vol. 4, pp. 528–590.
Drobnik, J. and *Rypáček, F.:* Soluble Synthetic Polymers in Biological Systems. Vol. 57, pp. 1–50.
Dröscher, M.: Solid State Extrusion of Semicrystalline Copolymers. Vol. 47, pp. 120–138.
Duduković, M. P. see *Kardos, J. L.:* Vol. 80, pp. 101–123.
Drzal, L. T.: The Interphase in Epoxy Composites. Vol. 75, pp. 1–32.
Dušek, K.: Network Formation in Curing of Epoxy Resins. Vol. 78, pp. 1–58.
Dušek, K. and *Prins, W.:* Structure and Elasticity of Non-Crystalline Polymer Networks. Vol. 6, pp. 1–102.
Dušek, K. see *Brosse, J.-C.:* Vol. 81, pp. 167–224.
Duncan, R. and *Kopeček, J.:* Soluble Synthetic Polymers as Potential Drug Carriers. Vol. 57, pp. 51–101.

Eastham, A. M.: Some Aspects of the Polymerization of Cyclic Ethers. Vol. 2, pp. 18–50.
Ehrlich, P. and *Mortimer, G. A.:* Fundamentals of the Free-Radical Polymerization of Ethylene. Vol. 7, pp. 386–448.
Eisenberg, A.: Ionic Forces in Polymers. Vol. 5, pp. 59–112.
Eiss, N. S. Jr. see *Yorkgitis, E. M.* Vol. 72, pp. 79–110.
Elias, H.-G., Bareiss, R. und *Watterson, J. G.:* Mittelwerte des Molekulargewichts und anderer Eigenschaften. Vol. 11, pp. 111–204.
Elsner, G., Riekel, Ch. and *Zachmann, H. G.:* Synchrotron Radiation Physics. Vol. 67, pp. 1–58.
Elyashevich, G. K.: Thermodynamics and Kinetics of Orientational Crystallization of Flexible-Chain Polymers. Vol. 43, pp. 207–246.
Enkelmann, V.: Structural Aspects of the Topochemical Polymerization of Diacetylenes. Vol. 63, pp. 91–136.
Entelis, S. G., Evreinov, V. V., Gorshkov, A. V.: Functionally and Molecular Weight Distribution of Telchelic Polymers. Vol. 76, pp. 129–175.
Epaillard, F. see *Brosse, J.-C.:* Vol. 81, pp. 167–224.
Evreinov, V. V. see *Entelis, S. G.* Vol. 76, pp. 129–175.

Ferruti, P. and *Barbucci, R.:* Linear Amino Polymers: Synthesis, Protonation and Complex Formation. Vol. 58, pp. 55–92.

Finkelmann, H. and *Rehage, G.:* Liquid Crystal Side-Chain Polymers. Vol. 60/61, pp. 99–172.
Fischer, H.: Freie Radikale während der Polymerisation, nachgewiesen und identifiziert durch Elektronenspinresonanz. Vol. 5, pp. 463–530.
Flory, P. J.: Molecular Theory of Liquid Crystals. Vol. 59, pp. 1–36.
Ford, W. T. and *Tomoi, M.:* Polymer-Supported Phase Transfer Catalysts Reaction Mechanisms. Vol. 55, pp. 49–104.
Fradet, A. and *Maréchal, E.:* Kinetics and Mechanisms of Polyesterifications. I. Reactions of Diols with Diacids. Vol. 43, pp. 51–144.
Franta, E. see Rempp, P.: Vol. 86, pp. 145–173.
Franz, G.: Polysaccharides in Pharmacy. Vol. 76, pp. 1–30.
Friedrich, K.: Crazes and Shear Bands in Semi-Crystalline Thermoplastics. Vol. 52/53, pp. 225–274.
Fujita, H.: Diffusion in Polymer-Diluent Systems. Vol. 3, pp. 1–47.
Funke, W.: Über die Strukturaufklärung vernetzter Makromoleküle, insbesondere vernetzter Polyesterharze, mit chemischen Methoden. Vol. 4, pp. 157–235.
Furukawa, H. see Kamon, T.: Vol. 80, pp. 173–202.

Gal'braikh, L. S. and *Rigovin, Z. A.:* Chemical Transformation of Cellulose. Vol. 14, pp. 87–130.
Galli, G. see Chiellini, E. Vol. 62, pp. 143–170.
Gallot, B. R. M.: Preparation and Study of Block Copolymers with Ordered Structures, Vol. 29, pp. 85–156.
Gandini, A.: The Behaviour of Furan Derivatives in Polymerization Reactions. Vol. 25, pp. 47–96.
Gandini, A. and *Cheradame, H.:* Cationic Polymerization. Initiation with Alkenyl Monomers. Vol. 34/35, pp. 1–289.
Geckeler, K., Pillai, V. N. R., and *Mutter, M.:* Applications of Soluble Polymeric Supports. Vol. 39. pp. 65–94.
Gerrens, H.: Kinetik der Emulsionspolymerisation. Vol. 1, pp. 234–328.
Ghiggino, K. P., Roberts, A. J. and *Phillips, D.:* Time-Resolved Fluorescence Techniques in Polymer and Biopolymer Studies. Vol. 40, pp. 69–167.
Gillham, J. K. see Aronhime, M. T.: Vol. 78, pp. 81–112.
Glöckner, G.: Analysis of Compositional and Structural Heterogeneitis of Polymer by Non-Exclusion HPLC. Vol. 79, pp. 159–214.
Godovsky, Y. K.: Thermomechanics of Polymers. Vol. 76, pp. 31–102.
Godovsky, Yu. K. and *Papkov, V. S.:* Thermotropic Mesophases in Element-Organic Polymers. Vol. 88, pp. 129–180.
Goethals, E. J.: The Formation of Cyclic Oligomers in the Cationic Polymerization of Heterocycles. Vol. 23, pp. 103–130.
Gorshkov, A. V. see Entelis, S. G. Vol. 76, 129–175.
Graessley, W. W.: The Etanglement Concept in Polymer Rheology. Vol. 16, pp. 1–179.
Graessley, W. W.: Entagled Linear, Branched and Network Polymer Systems. Molecular Theories. Vol. 47, pp. 67–117.
Grebowicz, J. see Wunderlich, B. Vol. 60/61, pp. 1–60.
Grebowicz, J. see Wunderlich, B.: Vol. 87, pp. 1–121.
Greschner, G. S.: Phase Distribution Chromatography. Possibilities and Limitations. Vol. 73/74, pp. 1–62.

Hagihara, N., Sonogashira, K. and *Takahashi, S.:* Linear Polymers Containing Transition Metals in the Main Chain. Vol. 41, pp. 149–179.
Hallpap, P., Bölke, M., and *Heublein, G.:* Elucidation of Cationic Polymerization Mechanisms by Means of Quantum Chemical Methods. Vol. 86, pp. 175–236.
Hasegawa, M.: Four-Center Photopolymerization in the Crystalline State. Vol. 42, pp. 1–49.
Hatano, M.: Induced Circular Dichroism in Biopolymer-Dye System. Vol. 77, pp. 1–121.
Hay, A. S.: Aromatic Polyethers. Vol. 4, pp. 496–527.
Hayakawa, R. and *Wada, Y.:* Piezoelectricity and Related Properties. of Polymer Films. Vol. 11, pp. 1–55.
Heidemann, E. and *Roth, W.:* Synthesis and Investigation of Collagen Model Peptides. Vol. 43, pp. 145–205.

Heinrich, G., Straube, E., and *Helmis, G.:* Rubber Elasticity of Polymer Networks: Theories. Vol. 84, pp. 33–87.
Heitz, W.: Polymeric Reagents. Polymer Design, Scope, and Limitations. Vol. 23, pp. 1–23.
Helfferich, F.: Ionenaustausch. Vol. 1, pp. 329–381.
Helmis, G. see Heinrich, G. Vol. 84, pp. 33–87.
Hendra, P. J.: Laser-Raman Spectra of Polymers. Vol. 6, pp. 151–169.
Hendrix, J.: Position Sensitive "X-ray Detectors". Vol. 67, pp. 59–98.
Henrici-Olivé, G. and *Olivé, S.:* Oligomerization of Ethylene with Soluble Transition-Metal Catalysts. pp. 496–577.
Henrici-Olivé, G. und *Olivé, S.:* Koordinative Polymerisation an löslichen Übergangsmetall-Katalysatoren. Vol. 6, pp. 421–472.
Henrici-Olivé, G. and *Olivé, S.:* Oligomerization of Ethylene with Soluble Transition-Metal Catalysts. Vol. 15, pp. 1–30.
Henrici-Olivé, G. and *Olivé, S.:* Molecular Interactions and Macroscopic Properties of Polyacrylonitrile and Model Substances. Vol. 32, pp. 123–152.
Henrici-Olivé, G. and *Olivé, S.:* The Chemistry of Carbon Fiber Formation from Polyacrylonitrile. Vol. 51, pp. 1–60.
Hermans, Jr., J., Lohr, D. and *Ferro, D.:* Treatment of the Folding and Unfolding of Protein Molecules in Solution According to a Lattic Model. Vol. 9, pp. 229–283.
Herz, J.-E. see Rempp, P.: Vol. 86, pp. 145–173.
Heublein, G. see Hallpap, P.: Vol. 86, pp. 175–236.
Higashimura, T. and *Sawamoto, M.:* Living Polymerization and Selective Dimerization: Two Extremes of the Polymer Synthesis by Cationic Polymerization. Vol. 62, pp. 49–94.
Higashimura, T. see Masuda, T.: Vol. 81, pp. 121–166.
Hlady, V. see Andrade, J. D.: Vol. 79, pp. 1–63.
Hoffman, A. S.: Ionizing Radiation and Gas Plasma (or Glow) Discharge Treatments for Preparation of Novel Polymeric Biomaterials. Vol. 57, pp. 141–157.
Holzmüller, W.: Molecular Mobility, Deformation and Relaxation Processes in Polymers. Vol. 26, pp. 1–62.
Hori, Y. see Kashiwabara, H.: Vol. 82, pp. 141–207.
Horie, K. and *Mita, I.:* Reactions and Photodynamics in Polymer Solids. Vol. 88, pp. 77–128.
Hutchison, J. and *Ledwith, A.:* Photoinitiation of Vinyl Polymerization by Aromatic Carbonyl Compounds. Vol. 14, pp. 49–86.

Iizuka, E.: Properties of Liquid Crystals of Polypeptides: with Stress on the Electromagnetic Orientation. Vol. 20, pp. 79–107.
Ikada, Y.: Characterization of Graft Copolymers. Vol. 29, pp. 47–84.
Ikada, Y.: Blood-Compatible Polymers. Vol. 57, pp. 103–140.
Imanishi, Y.: Synthese, Conformation, and Reactions of Cyclic Peptides. Vol. 20, pp. 1–77.
Inagaki, H.: Polymer Separation and Characterization by Thin-Layer Chromatography. Vol. 24, pp. 189–237.
Inoue, S.: Asymmetric Reactions of Synthetic Polypeptides. Vol. 21, pp. 77–106.
Ise, N.: Polymerizations under an Electric Field. Vol. 6, pp. 347–376.
Ise, N.: The Mean Activity Coefficient of Polyelectrolytes in Aqueous Solutions and Its Related Properties. Vol. 7, pp. 536–593.
Isihara, A.: Irreversible Processes in Solutions of Chain Polymers. Vol. 5, pp. 531–567.
Isihara, A.: Intramolecular Statistics of a Flexible Chain Molecule. Vol. 7, pp. 449–476.
Isihara, A. and *Guth, E.:* Theory of Dilute Macromolecular Solutions. Vol. 5, pp. 233–260.
Iwatsuki, S.: Polymerization of Quinodimethane Compounds. Vol. 58, pp. 93–120.

Janeschitz-Kriegl, H.: Flow Birefrigence of Elastico-Viscous Polymer Systems. Vol. 6, pp. 170–318.
Jenkins, R. and *Porter, R. S.:* Unpertubed Dimensions of Stereoregular Polymers. Vol. 36, pp. 1–20.
Jenngins, B. R.: Electro-Optic Methods for Characterizing Macromolecules in Dilute Solution. Vol. 22, pp. 61–81.
Johnston, D. S.: Macrozwitterion Polymerization. Vol. 42, pp. 51–106.

Kamachi, M.: Influence of Solvent on Free Radical Polymerization of Vinyl Compounds. Vol. 38, pp. 55–87.
Kamachi, M.: ESR Studies on Radical Polymerization. Vol. 82, pp. 207–277.
Kamide, K. and *Saito, M.:* Cellulose and Cellulose Derivatives: Recent Advances in Physical Chemistry. Vol. 83, pp. 1–57.
Kamon, T., Furukawa, H.: Curing Mechanisms and Mechanical Properties of Cured Epoxy Resins. Vol. 80, pp. 173–202.
Kaneda, A. see Kinjo, N.: Vol. 88, pp. 1–48.
Kaneko, M. and *Wöhrle, D.:* Polymer-Coated Electrodes: New Materials for Science and Industry. Vol. 84, pp, 141–228.
Kaneko, M. and *Yamada, A.:* Solar Energy Conversion by Functional Polymers. Vol. 55, pp. 1–48.
Kardos, J. L., Dudukovič, M. P., Dave, R.: Void Growth and Resin Transport During Processing of Thermosetting — Matrix Composites. Vol. 80, pp. 101–123.
Kashiwabara, H., Shimada, S., Hori, Y. and *Sakaguchi, M.:* ESR Application to Polymer Physics — Molecular Motion in Solid Matrix in which Free Radicals are Trapped. Vol. 82, pp. 141–207.
Kawabata, S. and *Kawai, H.:* Strain Energy Density Functions of Rubber Vulcanizates from Biaxial Extension. Vol. 24, pp. 89–124.
Keii, T., Doi, Y.: Synthesis of "Living" Polyolefins with Soluble Ziegler-Natta Catalysts and Application to Block Copolymerization. Vol. 73/74, pp. 201–248.
Kelley, F. N. see LeMay, J. D.: Vol. 78, pp. 113–148.
Kennedy, J. P. and *Chou, T.:* Poly(isobutylene-co-β-Pinene): A New Sulfur Vulcanizable, Ozone Resistant Elastomer by Cationic Isomerization Copolymerization. Vol. 21, pp. 1–39.
Kennedy, J. P. and *Delvaux, J. M.:* Synthesis, Characterization and Morphology of Poly(butadiene-g-Styrene). Vol. 38, pp. 141–163.
Kennedy, J. P. and *Gillham, J. K.:* Cationic Polymerization of Olefins with Alkylaluminium Initiators. Vol. 10, pp. 1–33.
Kennedy, J. P. and *Johnston, J. E.:* The Cationic Isomerization Polymerization of 3-Methyl-1-butene and 4-Methyl-1-pentene. Vol. 19, pp. 57–95.
Kennedy, J. P. and *Langer, Jr., A. W.:* Recent Advances in Cationic Polymerization. Vol. 3, pp. 508–580.
Kennedy, J. P. and *Otsu, T.:* Polymerization with Isomerization of Monomer Preceding Propagation. Vol. 7, pp. 369–385.
Kennedy, J. P. and *Rengachary, S.:* Correlation Between Cationic Model and Polymerization Reactions of Olefins. Vol. 14, pp. 1–48.
Kennedy, J. P. and *Trivedi, P. D.:* Cationic Olefin Polymerization Using Alkyl Halide — Alkyl-Aluminium Initiator Systems. I. Reactivity Studies. II. Molecular Weight Studies. Vol. 28, pp. 83–151.
Kennedy, J. P., Chang, V. S. C. and *Guyot, A.:* Carbocationic Synthesis and Characterization of Polyolefins with Si–H and Si–Cl Head Groups. Vol. 43, pp. 1–50.
Khoklov, A. R. and *Grosberg, A. Yu.:* Statistical Theory of Polymeric Lyotropic Liquid Crystals. Vol. 41, pp. 53–97.
Kinjo, N., Ogata, M., Nishi, K. and *Kaneda, A.:* Epoxy Molding Compounds as Encapsulation Materials for Microelectronic Devices. Vol. 88, pp. 1–48.
Kinloch, A. J.: Mechanics and Mechanisms of Fracture of Thermosetting Epoxy Polymers. Vol. 72, pp. 45–68.
Kissin, Yu. V.: Structures of Copolymers of High Olefins. Vol. 15, pp. 91–155.
Kitagawa, T. and *Miyazawa, T.:* Neutron Scattering and Normal Vibrations of Polymers. Vol. 9, pp. 335–414.
Kitamaru, R. and *Horii, F.:* NMR Approach to the Phase Structure of Linear Polyéthylene. Vol. 26, pp. 139–180.
Klosinski, P., Penczek, S.: Teichoic Acids and Their Models: Membrane Biopolymers with Polyphosphate Backbones. Synthesis, Structure and Properties. Vol. 79, pp. 139–157.
Kloosterboer, J. G.: Network Formation by Chain Crosslinking Photopolymerization and its Applications in Electronics. Vol. 84, pp. 1–62.
Knappe, W.: Wärmeleitung in Polymeren. Vol. 7, pp. 477–535.
Koenik, J. L. see Mertzel, E. Vol. 75, pp. 73–112.
Koenig, J. L.: Fourier Transforms Infrared Spectroscopy of Polymers, Vol. 54, pp. 87–154.

Kolařík, J.: Secondary Relaxations in Glassy Polymers: Hydrophilic Polymethacrylates and Polyacrylates: Vol. 46, pp. 119–161.
Kong, E. S.-W.: Physical Aging in Epoxy Matrices and Composites. Vol. 80, pp. 125–171.
Koningsveld, R.: Preparative and Analytical Aspects of Polymer Fractionation. Vol. 7.
Kosyanchuk, L. F. see Lipatov, Yu. S.: Vol. 88, pp. 49–76.
Kovacs, A. J.: Transition vitreuse dans les polymers amorphes. Etude phénoménologique. Vol. 3, pp. 394–507.
Krässig, H. A.: Graft Co-Polymerization of Cellulose and Its Derivatives. Vol. 4, pp. 111–156.
Kramer, E. J.: Microscopic and Molecular Fundamentals of Crazing. Vol. 52/53, pp. 1–56.
Kraus, G.: Reinforcement of Elastomers by Carbon Black. Vol. 8, pp. 155–237.
Kratochvila, J. see Mejzlik, J.: Vol. 81, pp. 83–120.
Kreutz, W. and *Welte, W.:* A General Theory for the Evaluation of X-Ray Diagrams of Biomembranes and Other Lamellar Systems. Vol. 30, pp. 161–225.
Krimm, S.: Infrared Spectra of High Polymers. Vol. 2, pp. 51–72.
Kuhn, W., Ramel, A., Walters, D. H. Ebner, G. and *Kuhn, H. J.:* The Production of Mechanical Energy from Different Forms of Chemical Energy with Homogeneous and Cross-Striated High Polymer Systems. Vol. 1, pp. 540–592.
Kunitake, T. and *Okahata, Y.:* Catalytic Hydrolysis by Synthetic Polymers. Vol. 20, pp. 159–221.
Kurata, M. and *Stockmayer, W. H.:* Intrinsic Viscosities and Unperturbed Dimensions of Long Chain Molecules. Vol. 3, pp. 196–312.

Ledwith, A. and *Sherrington, D. C.:* Stable Organic Cation Salts: Ion Pair Equilibria and Use in Cationic Polymerization. Vol. 19, pp. 1–56.
Ledwith, A. see Chiellini, E. Vol. 62, pp. 143–170.
Lee, C.-D. S. and *Daly, W. H.:* Mercaptan-Containing Polymers. Vol. 15, pp. 61–90.
Legeay, G. see Brosse, J.-C.: Vol. 81, pp. 167–224.
LeMay, J. D., Kelley, F. N.: Structure and Ultimate Properties of Epoxy Resins. Vol. 78, pp. 113–148.
Lesná, M. see Mejzlik, J.: Vol. 81, pp. 83–120.
Lindberg, J. J. and *Hortling, B.:* Cross Polarization — Magic Angle Spinning NMR Studies of Carbohydrates and Aromatic Polymers. Vol. 66, pp. 1–22.
Lipatov, Y. S.: Relaxation and Viscoelastic Properties of Heterogeneous Polymeric Compositions. Vol. 22, pp. 1–59.
Lipatov, Y. S.: The Iso-Free-Volume State and Glass Transitions in Amorphous Polymers: New Development of the Theory. Vol. 26, pp. 63–104.
Lipatov, Yu. S., Lipatova, T. E. and *Kosyanchuk, L. F.:* Synthesis and Structure of Macromolecular Topological Compounds. Vol. 88, pp. 49–76.
Lipatova, T. E.: Medical Polymer Adhesives. Vol. 79, pp. 65–93.
Lipatova, T. E. see Lipatov, Yu. S.: Vol. 88, pp. 49–76.
Lohse, F., Zweifel, H.: Photocrosslinking of Epoxy Resins. Vol. 78, pp. 59–80.
Luston, J. and *Vašš, F.:* Anionic Copolymerization of Cyclic Ethers with Cyclic Anhydrides. Vol. 56, pp. 91–133.

Madec, J.-P. and *Maréchal, E.:* Kinetics and Mechanisms of Polyesterifications. II. Reactions of Diacids with Diepoxides. Vol. 71, pp. 153–228.
Mano, E. B. and *Coutinho, F. M. B.:* Grafting on Polyamides. Vol. 19, pp. 97–116.
Maréchal, E. see Madec, J.-P. Vol. 71, pp. 153–228.
Mark, H. B., Jr. see Diaz, A. F.: Vol. 84, pp. 113–140.
Mark, J. E.: The Use of Model Polymer Networks to Elucidate Molecular Aspects of Rubberlike Elasticity. Vol. 44, pp. 1–26.
Mark, J. E. see Queslel, J. P. Vol. 71, pp. 229–248.
Maser, F., Bode, K., Pillai, V. N. R. and *Mutter, M.:* Conformational Studies on Model Peptides. Their Contribution to Synthetic, Structural and Functional Innovations on Proteins. Vol. 65, pp. 177–214.
Masuda, T. and *Higashimura, T.:* Polyacetylenes with Substituents: Their Synthesis and Properties. Vol. 81, pp. 121–166.

McGrath, J. E. see Yilgör, I.: Vol. 86, pp. 1–86.
McGrath, J. E. see Yorkgitis, E. M. Vol. 72, pp. 79–110.
McIntyre, J. E. see Dobb, M. G. Vol. 60/61, pp. 61–98.
Meerwall v., E. D.: Self-Diffusion in Polymer Systems. Measured with Field-Gradient Spin Echo NMR Methods, Vol. 54, pp. 1–29.
Mejzlik, J., Lesná, M. and *Kratochvila, J.:* Determination of the Number of Active Centers in Ziegler-Natta Polymerizations of Olefins. Vol. 81, pp. 83–120.
Mengoli, G.: Feasibility of Polymer Film Coating Through Electroinitiated Polymerization in Aqueous Medium. Vol. 33, pp. 1–31.
Mertzel, E., Koenik, J. L.: Application of FT-IR and NMR to Epoxy Resins. Vol. 75, pp. 73–112.
Meyerhoff, G.: Die viscosimetrische Molekulargewichtsbestimmung von Polymeren. Vol. 3, pp. 59–105.
Millich, F.: Rigid Rods and the Characterization of Polyisocyanides. Vol. 19, pp. 117–141.
Mita, I. see Horie, K.: Vol. 88, pp. 77–128.
Möller, M.: Cross Polarization — Magic Angle Sample Spinning NMR Studies. With Respect to the Rotational Isomeric States of Saturated Chain Molecules. Vol. 66, pp. 59–80.
Möller, M. see Wunderlich, B.: Vol. 87, pp. 1–121.
Morawetz, H.: Specific Ion Binding by Polyelectrolytes. Vol. 1, pp. 1–34.
Morgan, R. J.: Structure-Property Relations of Epoxies Used as Composite Matrices. Vol. 72, pp. 1–44.
Morin, B. P., Breusova, I. P. and *Rogovin, Z. A.:* Structural and Chemical Modifications of Cellulose by Graft Copolymerization. Vol. 42, pp. 139–166.
Mulvaney, J. E., Oversberger, C. C. and *Schiller, A. M.:* Anionic Polymerization. Vol. 3, pp. 106–138.

Nakase, Y., Kurijama, I. and *Odajima, A.:* Analysis of the Fine Structure of Poly(Oxymethylene) Prepared by Radiation-Induced Polymerization in the Solid State. Vol. 65, pp. 79–134.
Neuse, E.: Aromatic Polybenzimidazoles. Syntheses, Properties, and Applications. Vol. 47, pp. 1–42.
Nicolais, L. see Apicella, A. Vol. 72, pp. 69–78.
Nishi, K. see Kinjo, N.: Vol. 88, pp. 1–48.
Noristi, L. see Barbé, P. C.: Vol. 81, pp. 1–83.
Nuyken, O., Weidner, R.: Graft and Block Copolymers via Polymeric Azo Initiators. Vol. 73/74, pp. 145–200.

Ober, Ch. K., Jin, J.-I. and *Lenz, R. W.:* Liquid Crystal Polymers with Flexible Spacers in the Main Chain. Vol. 59, pp. 103–146.
Ogata, M. see Kinjo, N.: Vol. 88, pp. 1–48.
Okubo, T. and *Ise, N.:* Synthetic Polyelectrolytes as Models of Nucleic Acids and Esterases. Vol. 25, pp. 135–181.
Oleinik, E. F.: Epoxy-Aromatic Amine Networks in the Glassy State Structure and Properties. Vol. 80, pp. 49–99.
Osaki, K.: Viscoelastic Properties of Dilute Polymer Solutions. Vol. 12, pp. 1–64.
Osada, Y.: Conversion of Chemical Into Mechanical Energy by Synthetic Polymers (Chemomechanical Systems). Vol. 82, pp. 1–47.
Oster, G. and *Nishijima, Y.:* Fluorescence Methods in Polymer Science. Vol. 3, pp. 313–331.
Otsu, T. see Sato, T. Vol. 71, pp. 41–78.
Overberger, C. G. and *Moore, J. A.:* Ladder Polymers. Vol. 7, pp. 113–150.

Packirisamy, S. see Biswas, M. Vol. 70, pp. 71–118.
Papkov, S. P.: Liquid Crystalline Order in Solutions of Rigid-Chain Polymers. Vol. 59, pp. 75–102.
Papkov, V. S. see Godovsky, Yu. K.: Vol. 88, pp. 129–180.
Patat, F., Killmann, E. und *Schiebener, C.:* Die Absorption von Makromolekülen aus Lösung. Vol. 3, pp. 332–393.

Patterson, G. D.: Photon Correlation Spectroscopy of Bulk Polymers. Vol. 48, pp. 125–159.
Penczek, S., Kubisa, P. and *Matyjaszewski, K.:* Cationic Ring-Opening Polymerization of Heterocyclic Monomers. Vol. 37, pp. 1–149.
Penczek, S., Kubisa, P. and *Matyjaszewski, K.:* Cationic Ring-Opening Polymerization; 2. Synthetic Applications. Vol. 68/69, pp. 1–298.
Penczek, S. see Klosinski, P.: Vol. 79, pp. 139–157.
Peticolas, W. L.: Inelastic Laser Light Scattering from Biological and Synthetic Polymers. Vol. 9, pp. 285–333.
Petropoulos, J. H.: Membranes with Non-Homogeneous Sorption Properties. Vol. 64, pp. 85–134.
Pino, P.: Optically Active Addition Polymers. Vol. 4, pp. 393–456.
Pitha, J.: Physiological Activities of Synthetic Analogs of Polynucleotides. Vol. 50, pp. 1–16.
Platé, N. A. and *Noak, O. V.:* A Theoretical Consideration of the Kinetics and Statistics of Reactions of Functional Groups of Macromolecules. Vol. 31, pp. 133–173.
Platé, N. A., Valuev, L. I.: Heparin-Containing Polymeric Materials. Vol. 79, pp. 95–138.
Platé, N. A. see Shibaev, V. P. Vol. 60/61, pp. 173–252.
Plesch, P. H.: The Propagation Rate-Constants in Cationic Polymerisations. Vol. 8, pp. 137–154.
Porod, G.: Anwendung und Ergebnisse der Röntgenkleinwinkelstreuung in festen Hochpolymeren. Vol. 2, pp. 363–400.
Pospišil, J.: Transformations of Phenolic Antioxidants and the Role of Their Products in the Long-Term Properties of Polyolefins. Vol. 36, pp. 69–133.
Postelnek, W., Coleman, L. E., and *Lovelace, A. M.:* Fluorine-Containing Polymers. I. Fluorinated Vinyl Polymers with Functional Groups, Condensation Polymers, and Styrene Polymers. Vol. 1, pp. 75–113.

Queslel, J. P. and *Mark, J. E.:* Molecular Interpretation of the Moduli of Elastomeric Polymer Networks of Know Structure. Vol. 65, pp. 135–176.
Queslel, J. P. and *Mark, J. E.:* Swelling Equilibrium Studies of Elastomeric Network Structures. Vol. 71, pp. 229–248.

Rehage, G. see Finkelmann, H. Vol. 60/61, pp. 99–172.
Rempp, P. F. and *Franta, E.:* Macromonomers: Synthesis, Characterization and Applications. Vol. 58, pp. 1–54.
Rempp, P., Herz, J. and *Borchard, W.:* Model Networks. Vol. 26, pp. 107–137.
Rempp, P., Franta, E., and *Herz, J.-E.:* Macromolecular Engineering by Anionic Methods. Vol. 86, pp. 145–173.
Richards, R. W.: Small Angle Neutron Scattering from Block Copolymers. Vol. 71, pp. 1–40.
Rigbi, Z.: Reinforcement of Rubber by Carbon Black. Vol. 36, pp. 21–68.
Rigby, D. see Roe, R.-J.: Vol. 82, pp. 103–141.
Roe, R.-J. and *Rigby, D.:* Phase Relations and Miscibility in Polymer Blends Containing Copolymers. Vol. 82, pp. 103–141.
Rogovin, Z. A. and *Gabrielyan, G. A.:* Chemical Modifications of Fibre Forming Polymers and Copolymers of Acrylonitrile. Vol. 25, pp. 97–134.
Roha, M.: Ionic Factors in Steric Control. Vol. 4, pp. 353–392.
Roha, M.: The Chemistry of Coordinate Polymerization of Dienes. Vol. 1, pp. 512–539.
Ross-Murphy, S. B. see Clark, A. H.: Vol. 83, pp. 57–193.
Rostami, S. see Walsh, D. J. Vol. 70, pp. 119–170.
Rozengerk, v. A.: Kinetics, Thermodynamics and Mechanism of Reactions of Epoxy Oligomers with Amines. Vol. 75, pp. 113–166.
Rubinson, J. F. see Diaz, A. F.: Vol. 84, pp. 113–140.

Safford, G. J. and *Naumann, A. W.:* Low Frequency Motions in Polymers as Measured by Neutron Inelastic Scattering. Vol. 5, pp. 1–27.
Sakaguchi, M. see Kashiwabara, H.: Vol. 82, pp. 141–207.

Saito, M. see *Kamide, K.*: Vol. 83, pp. 1–57.
Sato, T. and *Otsu, T.:* Formation of Living Propagating Radicals in Microspheres and Their Use in the Synthesis of Block Copolymers. Vol. 71, pp. 41–78.
Sauer, J. A. and *Chen, C. C.:* Crazing and Fatigue Behavior in One and Two Phase Glassy Polymers. Vol. 52/53, pp. 169–224.
Sawamoto, M. see *Higashimura, T.* Vol. 62, pp. 49–94.
Schmidt, R. G., Bell, J. P.: Epoxy Adhesion to Metals. Vol. 75, pp. 33–72.
Schuerch, C.: The Chemical Synthesis and Properties of Polysaccharides of Biomedical Interest. Vol. 10, pp. 173–194.
Schulz, R. C. und *Kaiser, E.:* Synthese und Eigenschaften von optisch aktiven Polymeren. Vol. 4, pp. 236–315.
Seanor, D. A.: Charge Transfer in Polymers. Vol. 4, pp. 317–352.
Semerak, S. N. and *Frank, C. W.:* Photophysics of Excimer Formation in Aryl Vinyl Polymers, Vol. 54, pp. 31–85.
Seidl, J., Malinský, J., Dušek, K. und *Heitz, W.:* Makroporöse Styrol-Divinylbenzol-Copolymere und ihre Verwendung in der Chromatographie und zur Darstellung von Ionenaustauschern. Vol. 5, pp. 113–213.
Semjonow, V.: Schmelzviskositäten hochpolymerer Stoffe. Vol. 5, pp. 387–450.
Semlyen, J. A.: Ring-Chain Equilibria and the Conformations of Polymer Chains. Vol. 21, pp. 41–75.
Sen, A.: The Copolymerization of Carbon Monoxide with Olefins. Vol. 73/74, pp. 125–144.
Senturia, S. D., Sheppard, N. F. Jr.: Dielectric Analysis of Thermoset Cure. Vol. 80, pp. 1–47.
Sharkey, W. H.: Polymerizations Through the Carbon-Sulphur Double Bond. Vol. 17, pp. 73–103.
Sheppard, N. F. Jr. see *Senturia, S. D.*: Vol. 80, pp. 1–47.
Shibaev, V. P. and *Platé, N. A.:* Thermotropic Liquid-Crystalline Polymers with Mesogenic Side Groups. Vol. 60/61, pp. 173–252.
Shimada, S. see *Kashiwabara, H.*: Vol. 82, pp. 141–207.
Shimidzu, T.: Cooperative Actions in the Nucleophile-Containing Polymers. Vol. 23, pp. 55–102.
Shutov, F. A.: Foamed Polymers Based on Reactive Oligomers, Vol. 39, pp. 1–64.
Shutov, F. A.: Foamed Polymers. Cellular Structure and Properties. Vol. 51, pp. 155–218.
Shutov, F. A.: Syntactic Polymer Foams. Vol. 73/74, pp. 63–124.
Siesler, H. W.: Rheo-Optical Fourier-Transform Infrared Spectroscopy: Vibrational Spectra and Mechanical Properties of Polymers. Vol. 65, pp. 1–78.
Silvestri, G., Gambino, S., and *Filardo, G.:* Electrochemical Production of Initiators for Polymerization Processes. Vol. 38, pp. 27–54.
Sixl, H.: Spectroscopy of the Intermediate States of the Solid State Polymerization Reaction in Diacetylene Crystals. Vol. 63, pp. 49–90.
Slichter, W. P.: The Study of High Polymers by Nuclear Magnetic Resonance. Vol. 1, pp. 35–74.
Small, P. A.: Long-Chain Branching in Polymers. Vol. 18.
Smets, G.: Block and Graft Copolymers. Vol. 2, pp. 173–220.
Smets, G.: Photochromic Phenomena in the Solid Phase. Vol. 50, pp. 17–44.
Smets, G. see *Beylen, M. van*: Vol. 86, pp. 87–143.
Sohma, J. and *Sakaguchi, M.:* ESR Studies on Polymer Radicals Produced by Mechanical Destruction and Their Reactivity. Vol. 20, pp. 109–158.
Solaro, R. see *Chiellini, E.* Vol. 62, pp. 143–170.
Sotobayashi, H. und *Springer, J.:* Oligomere in verdünnten Lösungen. Vol. 6, pp. 473–548.
Soutif, J.-C. see *Brosse, J.-C.*: Vol. 81, pp. 167–224.
Sperati, C. A. and *Starkweather, Jr., H. W.:* Fluorine-Containing Polymers. II. Polytetrafluoroethylene. Vol. 2, pp. 465–495.
Spiertz, E. J. see *Vollenbroek, F. A.*: Vol. 84, pp. 85–112.
Spiess, H. W.: Deuteron NMR — A new Tool for Studying Chain Mobility and Orientation in Polymers. Vol. 66, pp. 23–58.
Sprung, M. M.: Recent Progress in Silicone Chemistry. I. Hydrolysis of Reactive Silane Intermediates, Vol. 2, pp. 442–464.
Stahl, E. and *Brüderle, V.:* Polymer Analysis by Thermofractography. Vol. 30, pp. 1–88.
Stannett, V. T., Koros, W. J., Paul, D. R., Lonsdale, H. K., and *Baker, R. W.:* Recent Advances in Membrane Science and Technology. Vol. 32, pp. 69–121.
Staverman, A. J.: Properties of Phantom Networks and Real Networks. Vol. 44, pp. 73–102.

Stauffer, D., Coniglio, A. and *Adam, M.:* Gelation and Critical Phenomena. Vol. 44, pp. 103–158.
Stille, J. K.: Diels-Alder Polymerization. Vol. 3, pp. 48–58.
Stolka, M. and *Pai, D.:* Polymers with Photoconductive Properties. Vol. 29, pp. 1–45.
Straube, E. see Heinrich, G.: Vol. 84, pp. 33–87.
Stuhrmann, H.: Resonance Scattering in Macromolecular Structure Research. Vol. 67, pp. 123–164.
Subramanian, R. V.: Electroinitiated Polymerization on Electrodes. Vol. 33, pp. 35–58.
Sumitomo, H. and *Hashimoto, K.:* Polyamides as Barrier Materials. Vol. 64, pp. 55–84.
Sumitomo, H. and *Okada, M.:* Ring-Opening Polymerization of Bicyclic Acetals, Oxalactone, and Oxalactam. Vol. 28, pp. 47–82.
Szegö, L.: Modified Polyethylene Terephthalate Fibers. Vol. 31, pp. 89–131.
Szwarc, M.: Termination of Anionic Polymerization. Vol. 2, pp. 275–306.
Szwarc, M.: The Kinetics and Mechanism of N-carboxy-α-amino-acid Anhydride (NCA) Polymerization to Poly-amino Acids. Vol. 4, pp. 1–65.
Szwarc, M.: Thermodynamics of Polymerization with Special Emphasis on Living Polymers. Vol. 4, pp. 457–495.
Szwarc, M.: Living Polymers and Mechanisms of Anionic Polymerization. Vol. 49, pp. 1–175.
Szwarc, M. see Beylen, M. van: Vol. 86, pp. 87–143.

Takahashi, A. and *Kawaguchi, M.:* The Structure of Macromolecules Adsorbed on Interfaces. Vol. 46, pp. 1–65.
Takemoto, K. and *Inaki, Y.:* Synthetic Nucleic Acid Analogs. Preparation and Interactions. Vol. 41, pp. 1–51.
Tani, H.: Stereospecific Polymerization of Aldehydes and Epoxides. Vol. 11, pp. 57–110.
Tate, B. E.: Polymerization of Itaconic Acid and Derivatives. Vol. 5, pp. 214–232.
Tazuke, S.: Photosensitized Charge Transfer Polymerization. Vol. 6, pp. 321–346.
Teramoto, A. and *Fujita, H.:* Conformation-dependent Properties of Synthetic Polypeptides in the Helix-Coil Transition Region. Vol. 18, pp. 65–149.
Theocaris, P. S.: The Mesophase and its Influence on the Mechanical Behvior of Composites. Vol. 66, pp. 149–188.
Thomas, W. M.: Mechanismus of Acrylonitrile Polymerization. Vol. 2, pp. 401–441.
Tieke, B.: Polymerization of Butadiene and Butadiyne (Diacetylene) Derivatives in Layer Structures. Vol. 71, pp. 79–152.
Tobolsky, A. V. and *DuPré, D. B.:* Macromolecular Relaxation in the Damped Torsional Oscillator and Statistical Segment Models. Vol. 6, pp. 103–127.
Tosi, C. and *Ciampelli, F.:* Applications of Infrared Spectroscopy to Ethylene-Propylene Copolymers. Vol. 12, pp. 87–130.
Tosi, C.: Sequence Distribution in Copolymers: Numerical Tables. Vol. 5, pp. 451–462.
Tran, C. see Yorkgitis, E. M. Vol. 72, pp. 79–110.
Tsuchida, E. and *Nishide, H.:* Polymer-Metal Complexes and Their Catalytic Activity. Vol. 24, pp. 1–87.
Tsuji, K.: ESR Study of Photodegradation of Polymers. Vol. 12, pp. 131–190.
Tsvetkov, V. and *Andreeva, L.:* Flow and Electric Birefringence in Rigid-Chain Polymer Solutions. Vol. 39, pp. 95–207.
Tuzar, Z., Kratochvil, P., and *Bohdanecký, M.:* Dilute Solution Properties of Aliphatic Polyamides. Vol. 30, pp. 117–159.

Uematsu, I. and *Uematsu, Y.:* Polypeptide Liquid Crystals. Vol. 59, pp. 37–74.

Valuev, L. I. see Platé, N. A.: Vol. 79, pp. 95–138.
Valvassori, A. and *Sartori, G.:* Present Status of the Multicomponent Copolymerization Theory. Vol. 5, pp. 28–58.
Vidal, A. see Donnet, J. B. Vol. 76, pp. 103–128.
Viovy, J. L. and *Monnerie, L.:* Fluorescence Anisotropy Technique Using Synchrotron Radiation as a Powerful Means for Studying the Orientation Correlation Functions of Polymer Chains. Vol. 67, pp. 99–122.

Voigt-Martin, I.: Use of Transmission Electron Microscopy to Obtain Quantitative Information About Polymers. Vol. 67, pp. 195–218.
Vollenbroek, F. A. and *Spiertz, E. J.:* Photoresist Systems for Microlithography. Vol. 84, pp. 85–112.
Voorn, M. J.: Phase Separation in Polymer Solutions. Vol. 1, pp. 192–233.

Walsh, D. J., Rostami, S.: The Miscibility of High Polymers: The Role of Specific Interactions. Vol. 70, pp. 119–170.
Ward, I. M.: Determination of Molecular Orientation by Spectroscopic Techniques. Vol. 66, pp. 81–116.
Ward, I. M.: The Preparation, Structure and Properties of Ultra-High Modulus Flexible Polymers. Vol. 70, pp. 1–70.
Weidner, R. see *Nuyken, O.:* Vol. 73/74, pp. 145–200.
Werber, F. X.: Polymerization of Olefins on Supported Catalysts. Vol. 1, pp. 180–191.
Wichterle, O., Šebenda, J., and *Králiček, J.:* The Anionic Polymerization of Caprolactam. Vol. 2, pp. 578–595.
Wilkes, G. L.: The Measurement of Molecular Orientation in Polymeric Solids. Vol. 8, pp. 91–136.
Wilkes, G. L. see Yorkgitis, E. M. Vol. 72, pp. 79–110.
Williams, G.: Molecular Aspects of Multiple Dielectric Relaxation Processes in Solid Polymers. Vol. 33, pp. 59–92.
Williams, J. G.: Applications of Linear Fracture Mechanics. Vol. 27, pp. 67–120.
Wöhrle, D.: Polymere aus Nitrilen. Vol. 10, pp. 35–107.
Wöhrle, D.: Polymer Square Planar Metal Chelates for Science and Industry. Synthesis, Properties and Applications. Vol. 50, pp. 45–134.
Wöhrle, D. see Kaneko, M.: Vol. 84, pp. 141–228.
Wolf, B. A.: Zur Thermodynamik der enthalpisch und der entropisch bedingten Entmischung von Polymerlösungen. Vol. 10, pp. 109–171.
Wolf, B. A. see Ballauff, M.: Vol. 84, pp. 1–31.
Wong, C. P.: Application of Polymer in Encapsulation of Electronic Parts. Vol. 84, pp. 63–84.
Woodward, A. E. and *Sauer, J. A.:* The Dynamic Mechanical Properties of High Polymers at Low Temperatures. Vol. 1, pp. 114–158.
Worsfold, D. J. see Beylen, M. van: Vol. 86, pp. 87–143.
Wunderlich, B.: Crystallization During Polymerization. Vol. 5, pp. 568–619.
Wunderlich, B. and *Baur, H.:* Heat Capacities of Linear High Polymers. Vol. 7, pp. 151–368.
Wunderlich, B. and *Grebowicz, J.:* Thermotropic Mesophases and Mesophase Transitions of Linear, Flexible Macromolecules. Vol. 60/61, pp. 1–60.
Wunderlich, B., Möller, M., Grebowicz, J. and *Baur, H.:* Conformational Motion and Disorder in Low and High Molecular Mass Crystals. Vol. 87, pp. 1–121.
Wrasidlo, W.: Thermal Analysis of Polymers. Vol. 13, pp. 1–99.

Yamashita, Y.: Random and Black Copolymers by Ring-Opening Polymerization. Vol. 28, pp. 1–46.
Yamazaki, N.: Electrolytically Initiated Polymerization. Vol. 6, pp. 377–400.
Yamazaki, N. and *Higashi, F.:* New Condensation Polymerizations by Means of Phosphorus Compounds. Vol. 38, pp. 1–25.
Yilgör, I. and *McGrath, J. E.:* Polysiloxane Containing Copolymers: A Survey of Recent Developments. Vol. 86, pp. 1–86.
Yokoyama, Y. and *Hall, H. K.:* Ring-Opening Polymerization of Atom-Bridged and Bond-Bridged Bicyclic Ethers, Acetals and Orthoesters. Vol. 42, pp. 107–138.
Yorkgitis, E. M., Eiss, N. S. Jr., Tran, C., Wilkes, G. L. and *McGrath, J. E.:* Siloxane-Modified Epoxy Resins. Vol. 72, pp. 79–110.
Yoshida, H. and *Hayashi, K.:* Initiation Process of Radiation-induced Ionic Polymerization as Studied by Electron Spin Resonance. Vol. 6, pp. 401–420.
Young, R. N., Quirk, R. P. and *Fetters, L. J.:* Anionic Polymerizations of Non-Polar Monomers Involving Lithium. Vol. 56, pp. 1–90.
Yuki, H. and *Hatada, K.:* Stereospecific Polymerization of Alpha-Substituted Acrylic Acid Esters. Vol. 31, pp. 1–45.

Zachmann, H. G.: Das Kristallisations- und Schmelzverhalten hochpolymerer Stoffe. Vol. 3, pp. 581–687.
Zaikov, G. E. see Aseeva, R. M. Vol. 70, pp. 171–230.
Zakharov, V. A., Bukatov, G. D., and *Yermakov, Y. I.:* On the Mechanism of Olifin Polymerization by Ziegler-Natta Catalysts. Vol. 51, pp. 61–100.
Zambelli, A. and *Tosi, C.:* Stereochemistry of Propylene Polymerization. Vol. 15, pp. 31–60.
Zucchini, U. and *Cecchin, G.:* Control of Molecular-Weight Distribution in Polyolefins Synthesized with Ziegler-Natta Catalytic Systems. Vol. 51, pp. 101–154.
Zweifel, H. see Lohse, F.: Vol. 78, pp. 59–80.

Subject Index

Accelerator 20
Aluminium corrosion 12
Anodic corrosion 12
Antimóny trioxide 23
Avrami equation parameters, crystallization 145, 150
— — —, mesophase formation 145, 146
Azocompounds 82, 85, 88, 105, 106

Benzophenone 93

Catenane 50
—, methods of synthesis 52
Cathodic corrosion 13
Cessation of polymerization due to vitrification 81
Chain scission 121
Chemically-controlled reaction 83, 95, 119
Coloring agents 25
Complexes of cycles with metal-containing compounds 66
Condis crystal 133
Corrosion, aluminium 12
—, anodic 12
—, cathodic 12
Coupling agents 24
Critical free volume 84, 110
Crosslinking 29, 121
Crystalline silica 22
Cyclizing imidization 81
Cyclolinear polyorganosiloxanes, chemical structure 154
— —, crystalline phases in 156
— —, heat of transition 160
— —, Kuhn segments of macromolecules 155
— —, mesomorphic state 157–159
— —, transtion parameters, effect of substituents 157, 160
Cyclourethane, molecular model 66

Degradation 120
Diffusion-controlled reaction 83, 90, 95, 119
Diisobutylsilandiol, mesophase 164, 165
2,5-Distyrylpyrazine 115
DSC data on linear polycyclosiloxanes 156
— — on poly(diethylsiloxane) 135, 149
— — on poly(dipropylsiloxanes) 135
— — on polyphosphazenes 168
Dynamic modulus 82

Elasticity of poly(diethylsiloxane) 150–153
Encapsulation 4
Epoxy molding compounds 4
— resin 18
Error, soft 16

Failure mode 9
Fick's law 11
Filler 22
—, spherical 36
First-order kinetics 86, 89, 95
— —, deviations 86, 89, 95
Flame retardants 23
Flexibilizers 20
Free volume 33
— — distribution 110
Freezing of reaction 83
Fulgide 105
Fused silica 22

Gel time 27
Glass transition in linear polycyclosiloxanes 157
— — in poly(diethylsiloxane) 134
— — in poly(dimethylsiloxane) 134
— — in polyphosphazenes 168
Glass transition temperature 83, 85, 97, 119, 123

Hardeners 19
Hermetic packages 3

Heterogeneity 83, 89
Hexakis(trimethylsiloxy)disiloxane, thermotropic phase 166
Homogeneous line width 112

Imidization, cyclizing 81
Inclusion compounds 63
Inokuti-Hirayama equation 91
Isotropization temperature, chain tacticity influence 158
– –, dependence on molecular weight 147, 148, 159
– –, effect of flexible spacers 158–160
– –, linear polycyclosiloxanes 157–159
– –, poly(diethylsiloxane) 136, 137, 139, 147, 148
– –, poly(dipropylsiloxane) 136, 139, 147, 148
– –, polyphosphazene 168

Knots 50

Ladder polymers see Polyorganosilsesquioxanes
Lamella, mesomorphic, growth rate, linear 145
–, –, in poly(diethylsiloxane) 143
–, –, in polytolylsilsesquioxane 163
Latent accelerator 20
Liquid crystalline polymer, polyphosphazene spacers 134
– – – –, polysiloxane spacers 134
– – state, polymers 132

Macrocycle aggregates 67
Macrocycles, metal-catenand 54
–, ordering by complexing 65
Metal-catenand macrocycles 54
Micro-Brownian motion 33
Microphase-separated structure 21
Mesomorphic state, classification 131
Mesophase coexistance with amorphous phase 141
– effect on crystallization 148–150, 175
– formation under stress 150–153
– in low molecular siloxanes 164–167
– kinetics of formation 143–146
– morphology 139–143, 162, 175–177
–, oligomer 153, 154
– structure 139–143, 165, 166, 173
Mobility 83
Molecular dynamics, poly(diethylsiloxane) 138
– –, polyphosphazenes 172, 173
– weight effect, transition parameters 146
– – influence, isotropization temperature 147 to 148, 159

– – –, mesophase morphology 162
Mold filling dynamics 39
– –, uniform 43
Molding compounds, epoxy 1 ff.
–, transfer 5
Mooney's viscosity 36

Norrish reactions 118

Occupied volume 33
Octaphenylcyclotetrasiloxane, mesomorphic state 166

Package crack 10
Packaging technology 3 ff.
Perrin equation 91
Phase transition, linear polycyclosiloxanes 156, 160
– –, poly(diethylsiloxane) 135, 137
– –, poly(dimethylsiloxane) 134
– –, poly(dipropylsiloxane) 135
– –, polyphosphazenes 168
Phosphorescence 93, 99
Photochemical hole burning (PHB) 111
Photodimerization 115
Photo-Fries rearrangement 9, 41, 85, 123
Plastic crystal 133
Poly(bistrifluoroethoxyphosphazene) crystalline phase structure 171
– dilatometry 169
– transition parameters 169, 170
Polycatenanes 57
Polycyclosiloxanes, glass transition in linear 157
–, DSC data on linear 156
–, isotropization temperature 157–159
Polydiacetylene 116
Poly(diethylsiloxane), crystalline phase structure 137
–, DSC data 135, 149
–, elasticity 150, 151
–, glass transition 134
–, isotropization temperature 136, 137, 139, 147, 148
–, mesomorphic state 139–145
–, miscibility with dimethylsiloxane oligomers 153, 154
–, molecular dynamics 138
–, phase transition parameters 136, 137, 154
–, rigidity of macromolecules 134
–, thermoelasticity 151–153
Poly(dimethylsiloxane), glass transition 134
–, phase transition parameters 136
–, rigidity of macromolecules 134
Poly(diphenylsiloxane), heat of fusion 134

Subject Index

Poly(diphenylsiloxane), melting point 134
Poly(dipropylsiloxane), DSC data 135
—, isotropization temperature 147, 148
—, melting point, molecular weight effect 147
—, phase transition parameters 136
—, rigidity of macromolecules 134
Polyimides 44, 114
Poly(methylmethacrylate) (PMMA) 79, 81, 88, 94
Polyorganosilsesquioxanes, chemical structure 160
—, formation of mesophase during polymerization 163
—, Kuhn segment of macromolecules 161
—, mesophase morphology 161–164
Poly/oxy(decamethylcyclohexasiloxane-2,8-diyl)/isotropization 159
Polyphosphazenes, chemical structure 167
—, crystalline structure 170–172
—, DSC data 168
—, glass transition 168
—, isotropization temperature 168
—, mesomorphic state 172–175
—, morphology 174–176
—, rotational barriers in macromolecules 167
—, transition parameters 168
Poly-p-phenylene sulfide 44
Polyrotaxane 57
—, molecular model 72
—, structure stability 58
—, X-ray diffraction 70
Poly(vinyl cinnamate) 83, 116
Post-curing 28
Pressure Cooker Test 11

Quinizarin 112

Radiation-induced polymerization 79
α-Ray 39

Relaxation map 85
Release agents 25
Resin rheological parameter analyzer 42
Rosin-Rammler equation 36
Rotaxane 50
—, methods of synthesis 52
Rubber elasticity 29

Soft error 16
Spherical filler 36
Spirobenzopyran 82, 85, 101, 105
Spiroindolinonaphthoxazine (SNOX) 90, 104
Stern-Volmer equation 91
Stress intensity factor 10
—, thermal 34, 35
Stretched exponential 90
Surface mount 16
Swarm complexes 69

Tetraphenylporphin 112
Thermal shrinkage 32
— stress 35
Thermoelasticity of poly(diethylsiloxane) 151–153
Topochemical reactions 79
Topological compounds 50
Transfer molding 5

Uniform mold filling 43

Viscocrystalline state 133, 134

X-ray data, linear polycyclosiloxanes 156, 159
— —, poly(diethylsiloxane) 134, 136, 140, 141
— —, polyorganosilsesquioxanes 162
— —, polyphosphazenes 172, 173

JUN 0 7 1989